Everyboay's
Comet

A Layman's Guide to
COMET HALE-BOPP

By Alan Hale

Cover Design by Ann Lowe
Cover Art by Kim Poor

High-Lonesome Books
Silver City, New Mexico 88062

ISBN # 0-944383-38-6
Library of Congress Catalog Card #96-77219

First Edition

High-Lonesome Books
P. O. Box 878
Silver City, NM 88062
(505) 388-3763

ACKNOWLEDGMENTS

Writing this book has been a labor of love. It would not have been possible if there had not been a fuzzy object in the constellation of Sagittarius on that July morning in 1995, so first and foremost I am grateful to that object for making its appearance in my telescope that night.

As far as people go, I'm sincerely grateful to Dutch and Cherie Salmon at High-Lonesome Books for having enough faith in me to take this project on. And thanks to Bill Shepard for steering me in their direction, and to Dick Borowski for making that contact possible.

The following individuals have helped enormously in providing me with various bits and pieces of information: Thomas Bopp, John Darden, Steve Edberg, John Fleck, Dan Green, David Levy, Bob Lunsford, Brian Marsden, Rob McNaught, Charles Morris, Dick Nelson, Warren Offutt, Jim Scotti, Gene Shoemaker, Russell Sipe, Dave Thomas, Helen and Harlon Towns, Ed Woten, and Don Yeomans.

The following were very helpful in tracking down the various copyrights and permissions so I could use the illustrations which help make this book what it is: Edwin Aguirre, Robert Burnham, Dennis di Cicco, Wendy Eck, Steve Edberg, Dave Finley, Helen Hart, George House, Peter Jedicke, Jeff Kanipe, Rob McNaught, Charles Morris, Sally MacGillivray, Dan Oppliger, Kim Poor, and Don Yeomans. And a big thanks to everyone who allowed me to use their photographs and artwork.

I hope I haven't left anybody out of either of the above two lists; if so, please accept my apologies.

I'm sincerely grateful to my proofreaders who took the time to go through the manuscript and point out things which needed correcting: Alice Buck, Ruth and Ray Genuske, Eva Hale, and Ruth Perkins (aka "Mom"). Any errors which are left belong to me.

A warm "thank you" to Janet Asimov and to Thomas Bopp for agreeing to share their thoughts via their Introductions. Likewise to Marilyn Haddrill, for taking one of the few decent photographs of me in existence.

I could go on and on listing those who have provided inspiration in my pursuits through the years. My father, Nile Hale, is the one who got me interested in astronomy in the first place, so any credit (or blame) should go to him.

Finally, I certainly must say "thank you" to my wife Eva and sons Zachary and Tyler, for in addition to having to put up with my regular nighttime activities, these past few months they have had to deal with their husband/father putting in endless hours in front of the computer writing and writing and writing... Their patience was endless.

I dedicate this book to two of my biggest sources of inspiration:

To Eva

May you finally get a chance to rest soon

and

To those twelve human beings who have walked on another world:

Neil Armstrong	Buzz Aldrin
Pete Conrad	Alan Bean
Alan Shepard	Ed Mitchell
David Scott	Jim Irwin
John Young	Charles Duke
Gene Cernan	Jack Schmitt

May there be many more who will follow in your footsteps someday

TABLE OF CONTENTS

THE COMET OF THE CENTURY.
Painting by Kim Poor, used with permission.

INTRODUCTION

During the wee hours on Sunday morning, July 23, 1995, nature gave me a present. That dim little fuzzy object that popped into my telescope's field of view that morning was a comet, but not just any comet; according to the calculations of its orbit it has the potential of being one of the brightest comets of the 20th Century, being a spectacular object in our nighttime skies during the spring months of 1997.

Throughout history, the appearance of a bright comet in the sky has excited the general populace on the Earth, and this object, Comet Hale-Bopp – named after myself and after Thomas Bopp of Glendale, Arizona, who discovered it the same night I did – should be no exception. Much of the excitement generated by previous comets contained a superstitious element, in that many people believed that the comets foretold future events (usually dire) or were harbingers of divine wrath, and so on. Even today, in our supposedly more enlightened society, these types of beliefs continue to occur. On the other hand, through the study of comets as scientific objects, we have been given tremendous insight into the processes by which the solar system formed, and by which it is shaped even today. In the course of this we have learned that there are times when comets should be feared, not as indicators of divine punishment, but as objects which could strike our Earth and threaten our civilization. (Comet Hale-Bopp, fortunately, poses no such threat).

My purposes in writing this book are threefold. First and foremost, I want to help you locate Comet Hale-Bopp when it is visible. Secondly, I want to combat all the misinformation that is sure to be flying around by 1997 – some of which has already started – and reassure you that this comet does not signify the end of the world, or any other such dire fate, but instead will be nothing more than a temporary (and hopefully spectacular) addition to our nighttime skies. And thirdly, by giving you some background about comets in general, what we have learned about them so far, and what we might hope to learn from Comet Hale-Bopp, I hope to give you a deeper understanding and appreciation of what you're going to be seeing in 1997.

I've broken this book down into three main sections. In Chapter 1 I talk about comets in general, some of the history involved in their study, and about the current scientific state of knowledge about them. There are plenty of good books on this subject available today, and I am not trying to compete with them; for the most part, Chapter 1 is a "quick-and-dirty" summary of the information that is available in these works. For the benefit of readers who might be interested in pursuing this subject further, I am providing a list of some of the better books available in Appendix A.

Chapter 2 discusses the early history of Comet Hale-Bopp, including the discovery stories of both Thomas Bopp and myself. I'll share with you the excitement that we astronomers began to feel as the true nature of the comet's orbit began to show itself, and I'll also present some of the early scientific results that have been obtained so far.

For most of you, Chapter 3 is the most important part; here I will tell you how to locate and follow Comet Hale-Bopp as it makes its journey across our skies during the coming year. I'll also try to tell you just what you might expect from the comet – and please try to keep in mind that there's still a lot of guesswork involved here, for reasons I'll go into – and I'll discuss some of the scientific observations that are planned during the comet's appearance.

When you're done, you should be able to go out into the nighttime sky, locate Comet Hale-Bopp, and understand the significance of what you're looking at. If, by this book, I have helped you attain such an appreciation of this object, and by extension have helped you feel a little closer to the universe around us, then I will have succeeded in all my goals. Happy comet viewing!

Alan Hale

PREFACE
by
Janet Asimov

One of the hallmarks of being human is the ability to use our minds in observing, thinking about, and exchanging ideas on the phenomena of nature.

The minds of our prehistoric ancestors were undoubtedly stimulated by the appearance of a great comet. There it was, a strange and awesome shape in the well-known territory of the sky, coming without warning from who knew where, and then disappearing into the unknown.

Legends grew about comets, and the impulse to create and remember these legends gave new impetus to the imagination.

There were also some people who were not content with legends, who wanted to know the truth. Eventually, the puzzle of comets may have helped promote the kind of human thinking that tries to find facts and explain them rationally – scientific reasoning.

Writers use imagination and, if they write about science, must be up on scientific thinking. As one of those writers, I must confess that I envy Alan Hale.

You see, I live in the light-polluted heart of Manhattan, where the moon and the brighter planets are about all the sky will usually reveal. From my bedroom window I managed to see a pale blob of light called Comet Hyakutake, but I missed the tail. And in the city one does not *discover* comets.

Dr. Hale is not only a scientist and a writer, but he lives where the sky is clear, and after twenty-five years of active observing (he started young), he discovered an important comet.

In writing about the comet that he and Thomas Bopp found on the same night, he has made this book much more than merely a guide to Comet Hale-Bopp.

With a personal touch and a sense of humor, Dr. Hale recounts the fascinating story of discovery and confirmation, plus the continued study of the intriguing changes in Hale-Bopp, which may be one of the "great comets" of history.

He also explains the science and history of comets, and gives practical advice about seeing them when they show up. His argument for scientific literacy is terrific, and his final sentence utterly compelling.

Before anyone asks, I'll reveal that Dr. Hale assures us that Comet Hale-Bopp is not one of those comets that will (according to recent headlines) hit Earth in the distant future.

This may disappoint people who expect humanity to be punished for ruining a perfectly good planet, but stick around because eventually one will hit. Learning all we can from each passing comet will help us prepare.

The best preparation, of course, is to get into a fail-safe situation, with colonies off Earth. Think what a great comet would be like from a dome on atmosphere-thin Mars, and its ice could be mined to replenish the colony's water supply!

As Comet Hale-Bopp nears Earth, but not too near, we can enjoy it. Comets are beautiful, intrinsically interesting, and immensely useful. Studying comets helps us understand our solar system, and it pays to understand the locality where you live.

Furthermore, our home base – Earth – would not be as livable as it is if it were not for comets.

When the solar system began 4.5 billion years ago, Earth was not prime real estate, being constantly bombarded with stuff left over from the formation of the planets. Some of the bombardment was by comets, which added more water and organic compounds to Earth. Scientists believe that our planet's organic chemistry – and eventually life itself – was speeded up by comets.

In every living organism there are probably molecules that were once in those comets. Each of us humans can say that we are part comet.

So, let's be grateful for dedicated people like the author of this book, who not only find comets but help us learn from them. And don't forget to watch for Hale-Bopp!

FOREWORD
by
Thomas Bopp

In mid-July 1994, the *Hubble Space Telescope* catapulted the world into a new awareness of space with the stunning images of the multiple impacts of fragments of Comet Shoemaker-Levy 9 into Jupiter's atmosphere. Unprecedented media activity held the public's interest as the dramatic events unfolded. Not since the much touted recovery of Halley's Comet in October 1982 has there been so much media attention focused on our night skies.

The return of Halley's Comet, however, was not the visual display many had hoped for. Indeed, not since Comet West in 1976 had there been a bright, naked eye comet that captured the attention of the general public. Then, on the night of July 22, 1995, Alan Hale and I independently co-discovered an 11th magnitude comet in the constellation Sagittarius. Little did either one of us know that night that this could be the comet that would end the drought.

Sunday morning, July 23rd, 1995, the morning after the discovery...having had a few fitful hours of sleep, the phone rang, and my wife Charlotte told me that someone was on the phone from the Harvard Smithsonian Astrophysical Observatory. I don't think my feet even touched the floor as I went for the telephone.

The caller identified himself as Dan Green and he wanted to know if I was Thomas Bopp and whether or not I had reported a comet. I replied that indeed I was and had and he congratulated me on the discovery. He told me that there was also an observer in New Mexico. He said that they believed it to be a new comet but that they would let me know. When I hung up the phone I invented a new dance around the kitchen floor. Four hours later Dan Green called again to verify some of the details of the discovery and told me that they would get back to me later.

Monday morning, July 24: Dan Green informed me that they already had multiple observations of the comet from Japan and Australia, and were well on the way to determining a preliminary orbit. He added that someone named Alan Hale in Cloudcroft, New Mexico, was the other observer who had also reported the comet.

Tuesday, July 25: Alan Hale telephoned me from Cloudcroft, New Mexico, and after we exchanged congratulations he told me that early orbital calculations indicated that the comet was a considerable distance from Earth, and for it to be visible in amateur telescopes at that distance it must be very large.

Wednesday, August 2nd: Alan called and with great excitement told me of a pre-discovery photograph of the comet by Rob McNaught in Australia, which allowed those who were involved in orbital calculation to determine that the comet was probably not making its first trip through the solar system and therefore had a much better chance of putting on a good visual display.

As our phone friendship grew we maintained a weekly discussion of the latest on Comet Hale-Bopp and talked of things to come. In September, 1995, Alan and his two sons and my father Frank Bopp and I were finally able to meet in Socorro, New Mexico at th Enchanted Skies Star Party.

Comet observing is an area of astronomy where amateurs can contribute useful information that can be used by professional astronomers to increase our understanding of the early solar system.

From pre-discovery to observing the comet at the peak of its display, the author motivates us to become personally involved in what may be a once in a lifetime event. Dr. Hale's comparison of the Great Comet of 1811 with Comet Hale-Bopp will whet your appetite for things to come.

Clear Skies,
Thomas Bopp

CHAPTER 1: WHAT IS A COMET?

When beggars die there are no comets seen.
The heavens themselves blaze forth the death of princes.
        William Shakespeare*

EARLY "SCIENCE" OF COMETS

Around 4:00 AM on Wednesday morning, March 3, 1976, my alarm clock started ringing. After shutting it off, and fumbling around for some clothes to put on, I sauntered out to my car in the pre-dawn darkness. After getting in, I fired up its ignition, and then took off in a westerly direction. My destination: White Sands National Monument, 15 miles west of my home in Alamogordo, New Mexico. My goal was to gather an observation of a comet that would soon be making its appearance in the morning sky. I had observed the comet twice in the evening sky a couple of weeks earlier, and the fact that I could see it then – at a time when it wasn't expected to be visible – suggested it might be a pretty nice sight by early March. The first two mornings in March were cloudy, and while it was clear on the 3rd, the Sacramento Mountains, which made up my eastern horizon, would cover the comet until the sky became too bright to make seeing it worthwhile. Thus, my trip to White Sands, away from the mounains.

I arrived at the parking lot there, and immediately noticed what appeared to be a bright searchlight beam extending up out of the mountains. I didn't think too much about this at first; I had a telescope and a camera to set up, and I busied myself by attending to these. A few minutes later, after I had set everything up, I still had about 10 minutes before I expected the comet to rise, and I turned my attention eastward toward the mountains. Again, I noticed the broad beam of light extending up from them, only this time I started to wonder just what it was I was seeing. Slowly, it dawned upon me that this was nothing other than the comet's tail, a fact which I verified a few minutes later when the comet's head rose. The sight of the brilliant comet, with its tail extending almost a quarter of the way up to the zenith (the point directly overhead) remains one of the most awe-inspiring I have seen in over thirty years of watching the sky.

The observation I just described is only one of innumerable sightings that have been made of comets throughout history. For the thousands of years that human beings have been aware of the night sky around them, from time to time they have been confronted with these majestic celestial wanderers and must

---

* *Julius Caesar* (Act II, Scene 2). Uttered by Calpurnia to her husband Caesar early in the morning of the Ides of March, and referring to the bright comet that appeared in 44 B.C. near the time of his assassination.

COMET WEST RISING OVER THE SACRAMENTO MOUNTAINS IN SOUTHERN NEW MEXICO. Photo taken from White Sands National Monument on March 3, 1976 by the author.

have wondered what they are. Ancient peoples, who were familiar with the so-called "fixed stars" as well as the "wandering stars" that we today know as the planets in our solar system, surely noticed these unexpected and imposing visitors when they appeared, and it isn't too surprising that these sights produced fear in the minds of the observers. Those who tried to find ways of foretelling the future from the happenings in the night sky can hardly be blamed for associating the bright comets that appeared with the catastrophic events that were bound to occur from time to time.

Many ancient peoples, understandably, thought comets to be supernatural objects, sent – more likely than not – as warnings of divine retribution against various shortcomings of humanity. On the other hand, the first attempt to explain, from a scientific point of view, just what a comet is seems to have come from the Greek philosopher Aristotle in the 4th Century B.C. He proposed that comets were "exhalations" of gases emanating from the ground into the upper atmosphere. In this way of thinking, comets, unlike the stars or planets, were not celestial objects, but instead were mere phenomena of our own air.

Before proceeding further with this, perhaps we should consider just what we see when we're looking at a comet. Of course we see the bright fuzzy "star," with the patch of light, or tail, attached to it. As we watch it over the course of an hour or so we should notice that it exhibits the same motion that the stars exhibit; i.e., the east-to-west path taken by all the stars as they rise, trek across the central sky, and then set. From night to night we may notice that the comet's position with respect to the stars changes slightly, and over a period of days or weeks we will probably see that physical features of the comet, such as its brightness and the length and direction of its tail, also appear to change.

It might be difficult – it certainly is to me – to see how all this could be caused by gases in the atmosphere, but such was the force of Aristotle's teachings that this assumption essentially remained unchallenged for several centuries. Alongside this there continued the association of comets with predictions of catastrophic events, and it was not at all unusual for a particularly dreadful event (for example, an epidemic, an assassination, a war – of which there were never any shortages) to be blamed on a comet that might be appearing at around the same time. In one famous incident, a comet that appeared in 1456 caused so much fear throughout the Mediterranean region that the Pope issued edicts requiring the people to pray for salvation from "the Devil, the Turk, and the Comet."

ONE VIEW OF A COMETARY ENCOUNTER. Cartoon from a 19th Century French magazine depicting the collision of a comet with the earth. Courtesy Don Yeomans.

One of the earliest "scientific" observations of a comet seems to have been made by the French mathematician Peter Apian in 1531. In his observations of a comet that appeared that year Apian noticed that the comet's tail always seemed to extend from the head in the direction opposite that of the sun. It seems, though, that this finding didn't make much of a splash in the overall thinking of the time, and thus people continued to think of the comets as nothing more than gases in the atmosphere that foretold doom.

HALLEY'S COMET IN 1531. These drawings by Peter Apian were used by him to deduce the fact that a comet's tail always points away from the sun. Courtesy Don Yeomans.

During the year 1577 an especially brilliant comet appeared in the sky, actually becoming bright enough to be visible during daylight. The top astronomer in the world at that time was a hot-tempered Danish man, Tycho Brahe (who reportedly lost part of his nose during a sword duel). Through careful observation Tycho – he is usually referred to by his first name – found that the comet didn't show any *parallax;* that is, it didn't change its position with respect to the stars if one viewed it from different locations on the earth. From this, Tycho deduced that the comet was at least four times further away than the moon; in other words, the comet was an astronomical object beyond the confines of Earth, not a phenomenon of the upper atmosphere.

(For those who want a demonstration of parallax, try the following exercise: hold a thumb up at arm's length and, while keeping one eye closed, line up the thumb with a distant object. Now, keeping the thumb stationary, close the open eye and open the other eye. The thumb will appear to "jump" with respect to the background object. It hasn't moved, of course; all that has happened is that it is being viewed from a different direction. If the same exercise is tried with the thumb closer to the face, one will notice a bigger "jump." The degree to which the thumb has changed its apparent position is called its *parallax*, and from this exercise one can see that the closer an object is, the larger is its parallax. This principle can be used to find the distances of objects in the solar system, by viewing them from different locations on the earth's surface. Similarly, by observing nearby stars at different times of the year – with the earth being in different parts of its orbit – and measuring their parallax with respect to the background stars, the distances to the nearby stars can also be measured).

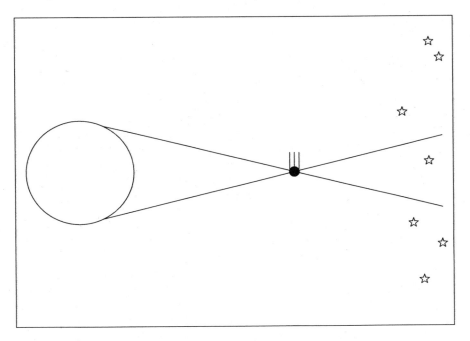

PARALLAX. From two widely separated points on the earth, a nearby object, such as a comet, will be displaced against the background of the distant stars.

Unlike Apian's observations, those of Tycho Brahe did start to be noticed by the scientists of the day, and although much of the general public continued to view comets with fear and trembling, cometary science began to advance. Indeed, the study of astronomy as a whole began to advance during this time; for example, the Italian astronomer Galileo Galilei began using his small telescope (adapted from an invention made by the Flemish spectacle-maker Hans Lippershey in 1609) to study the heavens in 1610 and, although we have no real records that he observed any comets with this, he did observe many other objects which eventually helped establish that the sun, not the Earth, is the center of the solar system. Nearly contemporary with Galileo was the Danish astronomer Johannes Kepler (who started out as an assistant to Tycho Brahe) who, through the analysis of thousands of observations of the planets, was able to show that the planets moved around the sun not in circles, but in elongated paths called *ellipses*. (In truth, the orbits of the planets are *almost* perfect circles, but not quite; it was this "not quite" that Kepler demonstrated.) Kepler also determined several other mathematical characteristics of the planets' orbits, and today we know that, for all intents and purposes, he was correct.

THE LEGACY OF EDMOND HALLEY

Certainly one of the greatest scientists of recent history was the English physicist Isaac Newton, who during the latter years of the 17th Century authored his most famous work, the *Principia*. In this book Newton established his famous laws of motion and his law of gravity, with which, among other things, it was finally possible to prove mathematically Kepler's findings.

Because Newton was somewhat of a loner and preferred to work independently, it was not especially easy for him to get his *Principia* published. Thus, he ended up being assisted by his good friend, the English astronomer Edmond Halley. Halley, who was somewhat of a "jack-of-all-trades" scientist and who was far more sociable than Newton, even came up with money from his own pocket to help get Newton's book published.

While Tycho Brahe had shown that comets were indeed celestial objects, the common thought in Newton's and Halley's day was that all comets were "one-time" visitors, and traveled more or less on straight line paths during their visit to the inner solar system. Halley, however, noting that with the math formulations in Newton's *Principia* it was possible to compute a planet's orbit, decided it might be interesting to apply the same procedure to comets and their orbits. He thus set about computing the orbits for 24 comets that had appeared in (his) relatively recent history. This was, of course, long before the days of calculators and computers, and since the calculations involved in determining orbits are quite laborious and lengthy, it understandably took Halley a long time to do this.

EDMOND HALLEY. Courtesy Steve Edberg.

One thing that Halley noticed when he had completed this set of calculations was that, while cometary orbits were not near-circles like those of the planets, their paths were not straight lines either. He found that the comets he was studying appeared to be traveling in very extended curved paths; the orbits in fact looked rather similar to open-ended curves called *parabolas*.

Another thing that Halley noticed was that the comets of 1531 (Peter Apian's object, incidentally), 1607 and 1682 (this last which he had himself observed) appeared to travel in very similar orbits. Furthermore, the two intervals separating these three comets were about 76 years in each case. Halley then made the interesting supposition that, instead of being three separate comets, these three objects were actually different returns of the same object, a comet that returns to the inner solar system every 76 years. Since 76 years after 1682 brings us to the year 1758, Halley made the bold suggestion that this comet would be visible again in that year. While he himself could not

reasonably expect to live long enough to see if his prediction would come true –
he would be 102 years old then – he uttered the rather nationalistic statement
that, if his prediction was verified, "...impartial posterity will not refuse to
acknowledge that this was first discovered by an Englishman."

As 1758 arrived several astronomers began serious searches for this comet,
but as the year rolled on and the searches were unsuccessful, it began to look as
though Halley might be wrong after all. But on Christmas night of that year, a
German amateur astronomer, Johann Palitzsch, discovered a dim comet which
turned out to be none other than the one that Halley had predicted. The comet
was actually closest to the sun during March of the following year; today we
know that the gravitational influences of the planets – Jupiter in particular – can
alter the orbit of a comet somewhat, and this is why Halley's prediction was a
year off.

The astronomers of that time, rightfully recognizing the genius of Halley in
making his prediction, decided it would be appropriate to name this comet in his
honor. Halley's comet, of course, has now become the most famous of all the
known comets, and since that time has made additional visits in 1835, 1910, and
(most recently) in 1986. In addition, several ancient historical comets have also
been identified as being earlier returns of Halley's comet; the earliest return that
has been identified is as far back as 240 B.C.

Once Halley's idea that the returns of comets could be predicted was so
spectacularly verified, the scientists of the late 18th Century began to examine
some of the other comets that were being seen then in an effort to see if any of
them were traveling in orbits which could allow similar predictions to be made.
In 1819 the German mathematician Johann Encke determined that three comets
observed in 1786, 1805, and 1818 were all in fact the same object, with an
orbital period as short as 3.3 years. Encke went on to predict that this object
would return in 1822, and when the comet was found then almost exactly where
Encke had said it would be, it was accordingly named in his honor. Even to this
day Encke's Comet has the shortest orbital period of any known comet, and to
date it has been observed at over 50 returns since Encke first made his
prediction. At this writing its most recent return was in early 1994, and it is next
expected to return in mid-1997.

Ever since Encke's day orbits have been calculated for numerous comets,
and the orbital periods have ranged from 4 years to several million years. Within
the past few decades astronomers have (rather arbitarily) defined a "periodic"
comet as one that has an orbital period of 200 years or less, and today almost
200 of these objects are known, with more being discovered every year. Over
120 of these have now been seen at two or more returns, and this tally also
grows with each passing year.

HOW COMETS ARE OBSERVED, DISCOVERED, DESIGNATED, AND NAMED

It should be kept in mind that most of these comets that I'm referring to are fairly dim objects, and usually require telescopes to be seen. Their numbers are surprisingly high; it may come as a surprise to most people to learn that, on any given clear night, about two or three comets are visible through an amateur astronomer's telescope, on the average, and to one of the large telescopes at professional observatories as many as one or two dozen might be visible. An experienced amateur astronomer could easily observe a dozen or more comets during an average year, and that number may well be three or four times higher for the large observatory instruments. Comets, then, are not rare events at all, although again I should stress that most of these are faint, dim objects that aren't of much interest except, of course, to those astronomers specifically interested in comets.

AN "AVERAGE" COMET. The typical comet appears as a "fuzzball" that moves against the background stars. These two photographs of such a comet were taken a couple of hours apart. NASA photograph, courtesy Table Mountain Observatory.

About once a year, on the average, there will appear a comet that will become bright enough to be visible with the naked eye. But keep in mind that, by "visible with the naked eye," I mean that a person with decently good vision, at a dark site away from city lights, and who knows exactly where to look, should be able to spot it without using binoculars or a telescope. The so-called "Great Comets," which are much more imposing – where the average person can simply look up in the night sky and see them without difficulty – are also much less frequent, and usually don't appear more often than about once every one to two decades. Until recently, the last such "Great Comet" to appear had been the object I observed from White Sands that morning back in 1976, and for the next few weeks after that, and some astronomers, arguing from the law of averages, had been pointing out that we were well overdue for such a visitor. The recent, brief, display of Comet Hyakutake, about which I'll say more later, has ended this drought, but in the meantime Comet Hale-Bopp most definitely possesses the potential for being another Great Comet, and for a much longer period of time than Hyakutake's display. Consequently, it is generating a lot of excitement among astronomers and lay people alike.

Although Halley's Comet and Encke's Comet are exceptions, for the past couple of centuries comets have traditionally been named for their discoverers. Thus, the comet I saw from White Sands was Comet West, named after Swiss astronomer Richard West who had discovered it several months earlier. Comet Hale-Bopp is named for its two discoverers, myself and Thomas Bopp.

Because of the obvious enjoyment in seeing one's name attached to a celestial object, throughout the past two centuries there has continued to flourish a dedicated group of astronomers who continuously scan the skies with their telescopes in hopes of discovering one or more comets. For many of these people searching for comets is an everyday practice, as they sweep the skies every clear night. Although discovering a comet is a matter of luck to some extent, if a person persists at this long enough, success will usually come at some point in time. Some individuals manage to discover several comets over the course of a lifetime, and every generation has had its standouts. These include the French astronomers Charles Messier in the late 18th Century and Jean Louis Pons in the early 19th; the Americans William Brooks, Lewis Swift, and Edward Barnard in the late 19th Century; American Leslie Peltier during the early 20th Century; and Antonin Mrkos in Czechoslovakia and the Japanese astronomers Minoru Honda, Kaoru Ikeya and Tsutomu Seki during the middle part of this century. The most prolific comet hunters today are Americans David Levy (with 8 comets – not counting those with which he assisted the Shoemakers, see below) and Don Machholz (with 9), and Australian astronomer William Bradfield (with 17).

The development of photography in the late 19th Century allowed astronomers to discover even fainter comets, and today a large percentage of the comets that are discovered are found this way. Until fairly recently, most such discoveries were accidental; i.e., a comet appeared on a photograph that an astronomer had taken for another purpose. But since about 1980 there have existed deliberate search programs at some of the larger observatories wherein professional astronomers have taken extensive series of photographs in an effort to detect comets (and asteroids). The most successful such program was conducted at Palomar Observatory in California by U.S. Geological Survey scientist Eugene Shoemaker and his wife Carolyn; from 1983 to 1994 they discovered no fewer than 32 new comets.

COMET HUNTER. *Howard Brewington, who lives near me in Cloudcroft, New Mexico. Howard is an active comet hunter, who has discovered four comets since 1989. He is diligently working on finding his fifth.*

It sometimes happens, as in the case with Comet Hale-Bopp, that a comet is discovered nearly simultaneously by different individuals. In such cases, it has been customary to name the comet after each of the discoverers (with the individual names separated by hyphens), up to a maximum number of three names. One can understand the reasoning behind this three-name limit by considering the fact that some comets remain hidden behind the sun and are not discovered until they are relatively bright naked-eye objects, at which point they are then simultaneously discovered by dozens of individuals. It would, of course, be ridiculous to have to recite two dozen names just to give the name of a comet; thus, the limit of three names. Even with this limit, there can be some rather tongue-twisting combinations put on these objects; recent decades have produced such oddly named comets as Honda-Mrkos-Pajdusakova, Bakharev-Macfarlane-Krienke, and Nakamura-Nishimura-Machholz. There has been a move afoot among some of the world's astronomers to limit the number of names on a comet to two, and perhaps we will see this within the relatively near future.

In addition to naming comets after their discoverers, astronomers have devised several schemes of assigning designations to comets. In part, this helps to distinguish between the various comets that might be discovered by the same individual. Until recently, the scheme that was used most was the assigning of a lower-case letter to the year of discovery in order to indicate when a particular comet was discovered; for example, **1994a** denotes the first comet that was discovered in 1994, **1994b** the second, and so on. Comet West, which I mentioned earlier, was the fourteenth comet discovered in 1975, and thus it was referred to as comet 1975n.

An alternative scheme used at the same time was the assigning of a designation to indicate when a comet passed through the *perihelion* point in its orbit, i.e., that point at which it is closest to the sun. (More about this in the next section.) This scheme involved the usage of Roman numerals to indicate the order of passage through perihelion; for example, Comet West was the sixth comet to pass perihelion in 1976, and thus was assigned the designation 1976 VI. Although these Roman numeral designations were generally referred to as the "permanent" designations of the comets (as opposed to the year/letter designations, which were generally considered "temporary"), there was an obvious disadvantage to using these in that they could not usefully be assigned until one or two years after the fact. Today it is not at all unusual for a comet to be discovered more than a year after it has passed perihelion, and if the Roman numeral designations were assigned too early, either the designations might have to be re-ordered after such a discovery (creating confusion), or the designations wouldn't necessarily reflect the precise order of perihelion passage (again creating some confusion). Despite the best efforts of astronomers, this did happen from time to time.

Partly because of this, and partly to eliminate the obvious redundancy in these schemes of dual designations, the International Astronomical Union – the worldwide community of astronomers which, among many other things, acts as a clearinghouse for announcing the discoveries of comets and other objects – introduced a new scheme at the beginning of 1995. The Roman numeral designations are being dispensed with entirely, and in place of the year/letter scheme for indicating a comet's discovery, astronomers are now using a year/letter/number scheme which can indicate rather precisely just when a comet is discovered. The "letter" indicates the particular half-month during which the discovery took place: A indicates the first half of January, B the second half of January, C the first half of February, and so on. For traditional reasons, the letter "I" is skipped; thus, the *first* half of May is denoted by the letter "J," the second half of May by "K," and so on.

The "number" indicates the order in which a comet was discovered during the half-month in question. For example, comet 1995 A1 is the first comet discovered during the first half of January 1995, 1995 B3 is the third comet discovered during the second half of January 1995, 1996 M2 is the second comet discovered during the second half of June 1996, and so on. A final part of this particular scheme includes the addition of a letter indicating whether or not a comet's orbital period is 200 years or less; if it is, the letter "P" is used (for example, comet P/1996 J1 means that the first comet discovered during the first half of May 1996 has an orbital period of less than 200 years). If not, the letter "C" is assigned. For example, comet C/1997 S4 means that the fourth comet discovered during the seond half of September 1997 has an orbital period greater than 200 years. Tom Bopp and I discovered our comet during the second half of July 1995, and it was the first comet found during that half-month; since our comet's orbital period is greater than 200 years, the full designation of the object is Comet C/1995 O1.

A FEW WORDS ABOUT COMETARY ORBITS

All the comets in our solar system – with one notable exception, which I'll discuss later – orbit around the sun, just like the planets do. But as Halley discovered, unlike the near-circular orbits of the planets, most comets travel in extremely elongated orbits, which bring them close to the sun for only a brief interval. While it might seem that the orbits of most comets are parabolas – i.e., orbits which are open-ended – it turns out that these are usually very elongated ellipses, and thus are closed (which means that the comets involved will return again some day). There are a few comets which do seem to have parabolic orbits, meaning they will never again return to the inner solar system; in fact, there are even some comets whose orbits are hyperbolas – which are curves even more "open" than parabolas – and these certainly will never return again. On the other hand, there are also a few comets known whose orbits are almost circular, like those of the planets. Thus, we see that cometary orbits can come in almost any shape.

In theory, the calculation of a comet's orbit is relatively straightforward, since it follows directly from Newton's law of gravity. The actual calculations themselves are rather tedious and difficult, although today with high-speed computers the calculations can be performed with little trouble. Once a comet is discovered, there are dedicated astronomers around the world who observe it and measure its position (with respect to the stars in its background), and it is from these observations that the orbit is calculated. At least three positions, spaced out over a couple of days, are required in order for an orbit to be calculated; as time goes by and more and more positions are obtained, the comet's "true" orbit becomes better and better defined.

Once the orbit is calculated, it is defined by a series of six quantities called "elements." Collectively, these elements describe the size, shape, and orientation of the comet's orbit. From these quantities, it is possible to calculate what is called an *ephemeris*, a table which gives the comet's position at future dates. Aided by an ephemeris, astronomers and other people can go out and observe the comet.

The most important elements of a comet's orbit are those which are associated with its *perihelion* point, that point at which it is closest to the sun. (The word comes from the Latin *peri*, meaning "near," and the Greek *helios*, meaning "sun." The point in the orbit farthest from the sun is called "aphelion.") Together, these elements describe when the comet is at perihelion, and how far away from the sun that perihelion point is. Typically, a comet is brightest when it is near perihelion, and the closer it is to the sun, the brighter this is; thus, these two elements can give a rough idea of when a comet will be visible and how bright it might get.

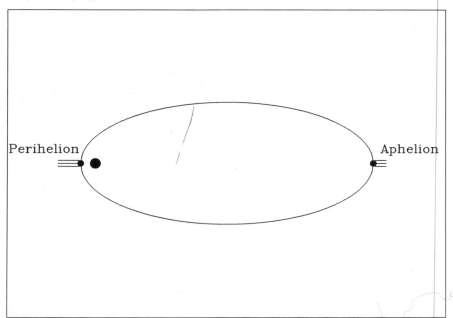

Perihelion   Aphelion

A TYPICAL COMETARY ORBIT. Some of the most important points are indicated (see text). The sun is indicated by the large black dot near the comet's perihelion.

When it comes to discussing distances in the solar system, most astronomers, instead of using miles (or kilometers), use what is called the *Astronomical Unit*

(AU). 1 AU is defined as the average distance between the Earth and the sun, or 93 million miles (or 149 million kilometers). Thus, a comet with a perihelion distance of 1 AU comes just as close to the sun as the earth gets, one whose perihelion point is 0.5 AU passes half the distance that the earth does, and so on. Halley's Comet has a perihelion distance of 0.59 AU, or 55 million miles (87 million kilometers). Among comets that have been seen so far, the perihelion distances have ranged from 0.005 AU (which is less than the radius of the sun – in other words, this particular comet actually hit the sun) to 8.54 AU, which is almost as far out as the orbit of Saturn (9.5 AU).

One other important point to remember about cometary orbits is that most of them are not in the same plane as the earth's orbit; usually they are tilted with respect to the Earth's orbit by some amount. The amount of this tilt – usually given in degrees – is called the *inclination*, and this is one of the six orbital elements. Those comets with a small inclination travel in orbits that are not tilted too much with respect to the Earth's orbit, whereas the higher the inclination angle, the higher is this tilt. There have been some comets known – and Hale-Bopp is one of these, incidentally – that travel in orbits with an inclination near 90°; in other words, their orbit is perpendicular to the earth's. It is even possible for an orbit to have an inclination greater than 90°; such an orbit is called *retrograde* (from the Latin word *retro*, meaning "backward") and any comet in such an orbit will travel around the sun in the direction opposite that of the earth. A comet with an inclination of exactly 180° would be traveling in the exact same plane as the Earth's orbit but would be moving in the opposite direction.

EXACTLY WHAT ARE COMETS?

Before answering the question of exactly what a comet is, one should perhaps consider what a comet looks like. The word "comet" itself comes from the Greek word *kome*, meaning "hair," and indeed, to the ancient peoples comets were considered more or less as "hairy stars." Usually, the "head" of a comet will appear as a fuzzy star-like object, and this head is called the "coma" (from the same Greek root word). The other main part of a comet is of course the tail which, as Apian found with Halley's Comet in the 16th Century, is almost always directed away from the sun. This means that as the comet approaches the sun, the coma will lead the tail, but as it leaves the sun, the tail will actually lead the head. I'll discuss the reasons for this a bit later.

Perhaps I should point out right here that not all comets develop significant tails; in fact, the majority of comets that are visible in the skies during any given night or year do *not* possess tails or, if they do, the tails are short and insignificant. Usually, it is just the coma itself that is seen when an astronomer looks at a comet. For the most part, a comet doesn't develop much in the way of a tail unless it gets pretty close to the sun, and many comets just never get close enough to the sun to develop one.

PARTS OF A COMET. Courtesy Steve Edberg.

As far as sizes go, there's been quite a bit of variation seen in the comets that have been observed so far, but the "typical" cometary coma is approximately 10,000 miles across. However, as I'll discuss a little later, this is not a solid object. The largest coma that has been seen so far measured almost a million miles across – actually larger than the sun (865,000 miles). The lengths of comets' tails can vary dramatically from comet to comet, and often from time to time for the same comet, but typically they extend a few million miles. Again, there can be extreme cases; the longest comet's tail that has ever been seen extended almost 2 AU in length.

Within the coma of most comets is a small bright area which is called – somewhat unimaginatively – the "central condensation." It is tempting to think of this as the comet's true center, or "nucleus," but this is actually not the case. What one sees when one looks at a comet's central condensation is a dense cloud of material surrounding the actual nucleus; the nucleus itself is hidden from view by all this material.

Because of this, the true nature of a cometary nucleus remained a mystery for a long time, and as the 20th Century progressed various ideas were put forth as to what it looked like and how it behaved. Finally, in 1950 an American astronomer, Fred Whipple, proposed the idea that a comet's nucleus is a "dirty snowball" composed of a series of ices – primarily ordinary water ice, but also including carbon dioxide ice ("dry ice"), carbon monoxide ice, and other similar substances – intermixed with a healthy proportion of interplanetary dust. According to Whipple's idea, as this "dirty snowball" starts to experience heat from the sun, the various ices start to *sublimate;* i.e., change directly from a solid to a gas, which can be accomplished in the near-vacuum of space. As this happens the dust is carried along with the gas, producing the coma and tail we eventually see.

While Whipple's idea made sense and seemed to do a pretty good job of explaining the different phenomena seen in various comets, there was no way to verify it without seeing an actual cometary nucleus. This finally happened in 1986, when the European Space Agency's *Giotto* spacecraft passed within a few hundred miles of the nucleus of Halley's Comet. *Giotto's* cameras recorded images of a dark, peanut-shaped object about 10 miles long by 5 miles wide, from which several "fountains" of material were seen to be erupting. This type of scenario was almost exactly what Fred Whipple had predicted some 36 years earlier, and has vindicated his "dirty snowball" theory.

The picture we have, then, of a comet is something like this: far out in space it is a cold, inert "dirty snowball" perhaps a few miles across. As it starts to move in toward the sun and starts experiencing heat, the various ices start to sublimate. Initially these include carbon monoxide, carbon dioxide, and others, and then, at about 3½ AU from the sun, the water ice – the predominant constituent – also begins to sublimate. The gas, together with the dust that is imbedded in the ice, is ejected from the nucleus through geyser-like eruptions (quite similar, in fact, to Old Faithful and its cousins in Yellowstone National Park), and this in turn creates a large cloud of dust and gas – the coma – around the nucleus. If the comet gets close enough to the sun, the solar wind – an energetic stream of particles constantly "blowing" off the sun – blows some of this gas and dust behind the coma to form a tail. (This is the reason a comet's tail is directed away from the sun.) Quite often, two tails are formed: one composed of gas, which has often been "ionized," or electrically charged, by the solar wind, and the other one made up of the dust.

As the comet recedes from the sun, the reverse happens. The further away it gets, the less heat it receives, and the less gas and dust are ejected from the nucleus. Eventually, the nucleus "shuts down" and the gas and dust in the coma and tail disperse away, eventually leaving nothing but the cold, inert "dirty snowball," which will remain in that state until it once again approaches the sun perhaps decades or centuries later.

THE NUCLEUS OF HALLEY'S COMET. Image taken by the Halley Multicolour Camera on board the European Space Agency's Giotto spacecraft during its encounter with Halley's Comet on March 13, 1986. Material can be seen erupting off the surface; this produces the coma and, eventually, the tail. Copyright 1986 by Max-Planck-Institut für Aeronomie, Lindau/Harz, FRG. Image supplied courtesy H. Uwe Keller.

I should clarify one point before moving on. Once the gas in a comet's nucleus gets heated by the sun, various chemical reactions occur, producing additional types of gases that will appear in both the coma and the tail. Some of these gases, notably Cyanogen, are what we might think of as "poisonous," and this has in the past led to fears that our atmosphere might be poisoned if it should ever get brushed by a comet's tail. As it turns out, even though a comet's tail may look pretty substantial in our night sky, the material in it is actually quite rarefied; it in fact is closer to a true "vacuum" than anything we can produce in our laboratories. It's been estimated that there is more gas in a matchbox full of air than there is in several cubic miles of a comet's tail. Thus, there is no danger of being poisoned if the earth should ever pass through a tail. This, in fact, has happened on occasion, most notably with Halley's Comet in 1910; despite some panic that spread before the event itself, there were no effects of any kind upon the earth's atmosphere.

WHERE DO COMETS COME FROM?

The question as to where the comets come from is inextricably tied in with the question of how the solar system formed. Although the picture is not complete yet – after all, there was no one around to take records – by studying young stars in space and by applying the known laws of physics to intensive computer calculations astronomers have, over the past 20 years or so, developed a fairly comprehensive picture of how this occurred.

Some 4.6 billion years ago, one small portion of a large cloud of dust and gas – we see several of these clouds scattered across the galaxy – started to collapse in on itself. Much of this material settled into a massive object near the center, and this eventually became the sun. At the same time, some of this material settled onto a wide thin disk that circled the young sun's equator. Over hundreds of thousands of years, this material began to stick together, until finally objects typically a few miles across had formed. After several more hundreds of thousands of years, collisions between these objects, called "planetesimals," began to produce the planets that we know today. Not all of the planetesimals were used up in forming the planets, though. Some, thanks to the gravity of some of the forming planets, were ejected from the solar system altogether, while others just simply got left behind. Those planetesimals that were made up primarily of "hard" materials such as silicates and metals are what we call the asteroids today; those that were made up primarily of substances such as ices are what we now know as the comets. In a very real sense, then, the comets and asteroids are "leftovers" from the formation of the solar system.

The asteroids, being rather resistant to the sun's heat, are able to reside in the inner solar system. Thanks primarily to Jupiter's gravity, most, though by no means all, of them reside in the main "asteroid belt" between the orbits of Mars and Jupiter. On the other hand, if the comets were to remain in this area they would eventually sublimate away entirely, and thus there are very few comets which remain permanently in the asteroid belt region. Most of the comets, then, reside in the cold outer regions of the solar system, beyond the influence of the sun's heat.

During the mid-20th Century a Dutch astronomer, Jan Oort, performed a study of cometary orbits to see if he could determine their origin and history. In 1950 – at about the same time that Fred Whipple was proposing his "dirty snowball" idea – Oort published a theory wherein he concluded that the comets originate in a vast spherical "cloud" that extends toward the furthest reaches of the solar system. This so-called "Oort Cloud" may extend out several tens of thousands of Astronomical Units – a significant fraction of the distance to the nearest star – and may contain several trillion comets.

A TYPICAL ASTEROID. The asteroid Gaspra, photographed by the Galileo *spacecraft on October 6, 1991. The entire asteroid is about 20 miles across. NASA photograph.*

For obvious reasons, it is impossible, with our present state of technology, to verify the existence of the Oort Cloud directly; the comets at that distance are simply far too dim to be visible with any telescopes we might have. But, from studying the orbits of numerous long-period comets, and calculating the distances of their aphelion points, astronomers can reproduce and extend Oort's calculations, and thus infer the existence of the Oort Cloud from this circumstantial evidence.

The vast majority of the comets that started out in the Oort Cloud are still there, in the cosmic deep-freeze many times farther than the orbit of Pluto. Occasionally, perhaps due to the gravitational influence of a nearby star that the sun might pass by on its wanderings around the galaxy, some of these comets might get "kicked in" toward the inner solar system. The journey is a long one:

the comets that we see from the Oort Cloud today probably started their trip inward millions of years ago.

If the comet and the sun were the only objects in the solar system, the comet would pass its perihelion, then head back out toward the Oort Cloud, not to make its next trip inward for another several million years or more. But, of course, there are the planets in the inner solar system, and each of these exerts its own gravitational influences on the comet's orbit. While this causes some comets to get placed on hyperbolic orbits, and thus ejected permanently from the solar system, many others get placed into orbits which bring them back to the sun's vicinity much more often. The comet's orbit is modified somewhat each time it returns to perihelion, and after several such visits it may find itself in an orbit that returns it to the sun after only centuries, or even decades.

The biggest culprit in this whole scenario is Jupiter, the largest planet, and correspondingly the one whose gravity exerts the most influence over comets' orbits. During the course of its many visits to the inner solar system a comet is bound to pass close to Jupiter sooner or later, and if the conditions are right, the comet may get "captured" by Jupiter's gravity into a very short-period orbit. The comets which have been so captured are usually collectively referred to as Jupiter's "family" of comets, and today over 120 members of this "family" are known, with additional members being discovered each year. The typical member of Jupiter's family has an orbital period of six to eight years, and for the few months of that cycle during which it is close to the sun it can be observed from here on the earth. Although very few of these objects get very bright, on any given night one or two of them are usually visible in moderate-sized backyard telescopes.

At about the same time that Oort was proposing his idea of the Oort Cloud, an American astronomer, Gerard Kuiper, advanced the idea that an inner zone of comets might exist beyond Neptune's and Pluto's orbits. At that time there didn't seem to be much evidence for the existence of this "inner Oort Cloud," and consequently it received very little attention from the rest of the astronomical community. But by the late 1980s, some astronomers began to argue that the Oort Cloud, as currently envisioned, just couldn't quite account for the large number of short-period comets seen in the inner solar system today, and they revived the idea of this so-called "Kuiper Belt" as an explanation.

One exciting aspect of this idea was that, with the telescope technology that was becoming available at the time, it seemed it might actually be possible to detect objects within the Kuiper Belt, and searches for these objects began in the early 1990s. The initial searches were unsuccessful, but then in August 1992 astronomers David Jewitt and Jane Luu announced that they had discovered an extremely dim slow-moving object from Mauna Kea Observatory in Hawaii. Several similar objects have been discovered since then, and studies have shown that all of them are traveling in near-circular orbits beyond Neptune, exactly where the Kuiper Belt is presumed to exist.

A FORMING SOLAR SYSTEM? Image of the star Beta Pictoris, taken by Richard Terrile and Brad Smith at Las Campañas Observatory in Chile. The star itself has been blocked out, and a thin disk of material accompanying it can be seen. The disk is composed of dust and small objects like planetesimals, and probably represents an early stage in the life of a planetary system. The disk's distance from Beta Pictoris is similar to the distance of the Kuiper Belt from our own sun. NASA photograph, courtesy Jet Propulsion Laboratory and Richard Terrile.

The exact nature of these objects is still quite problematical; their brightness suggests that they are a couple of hundred miles across, about the size of a moderately-sized asteroid. Comets, which would be expected to be much smaller, would also be much more difficult to detect, but quite recently a team of astronomers led by Anita Cochran at the University of Texas announced that, in a series of photographs taken with the *Hubble Space Telescope* during August 1994, they found evidence for over 50 extremely faint objects with the expected motion of objects within the Kuiper Belt. These objects have about the same brightness that Halley's Comet would be expected to have at that distance, which suggests that we are indeed seeing objects that may be the "Halley's Comets" of future millenia. The area of sky that was imaged with the *Hubble Space Telescope* was actually quite small, and the astronomers involved estimate that there may be as many as 100 million of these objects orbiting the sun within the Kuiper Belt.

Incidentally, some astronomers have determined that if one could examine our solar system from outside from a distance – say, from a nearby star – with a telescope sensitive to infrared radiation, the Kuiper Belt would be one of the brightest features visible. Some of the nearby stars do exhibit such a feature when examined with infrared telescopes, and this perhaps could be taken as evidence that other solar systems like ours are common throughout the universe.

COMETS AND METEORS

The material that is ejected from the comet's nucleus as it rounds the sun never returns to it, of course. The gas eventually breaks down to its separate molecules and disperses into interplanetary space. Meanwhile, each grain of dust settles into its own orbit around the sun, an orbit that is essentially the same as that of the comet. Eventually, the dust spreads out all along the comet's orbit.

If the comet's orbit should happen to pass close to the earth's orbit, there will come a time each year when the earth will intersect this stream of traveling dust. When this happens, a collision between the earth and the dust grain results, but because the dust grains are so tiny – typically just a micron (one-thousandth of a millimeter) across – they burn up in the atmosphere as a result of friction. (The typical speeds with which these dust grains hit the atmosphere is something like 20 to 40 miles per second.) Here on Earth we see these as tiny steaks of light in the nighttime sky; these are what we call "meteors," or sometimes "shooting stars."

Since such dust grains are scattered more or less everywhere throughout the inner solar system, several meteors can be seen during any given clear night. But when the earth intersects the stream of dust in a comet's orbit, the number of meteors that appear will increase and, moreover, they will appear to come from the same point in the sky. These events are called "meteor showers," and are usually named after the constellation from which they appear to originate. One of the strongest and most famous meteor showers is the Perseids, which recurs each year about August 12, and which produces from 60 to 100 meteors per hour that appear to emanate from the constellation Perseus.

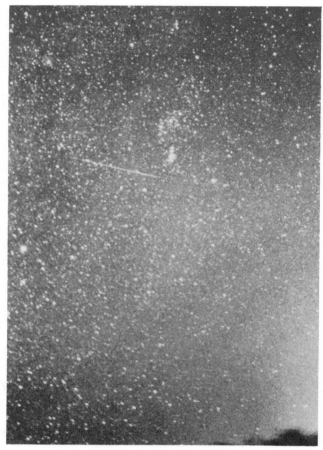

A METEOR STREAKS THROUGH ORION. Photograph by Robert Lunsford, used with permission.

This particular relationship between comets and meteor showers was first shown in the late 1860s by the Italian astronomer Giovanni Schiaparelli, who noted that the Perseids shared the same orbital characteristics as Comet Swift-Tuttle, which had appeared earlier that decade in 1862. Since that time, numerous other comet/meteor shower pairings have been identified; for example, Halley's Comet is associated with both the Eta Aquarid meteor shower (so called to distinguish it from other meteor showers that originate from Aquarius throughout the year) that peaks around May 3, and the Orionids, which recur each year around October 21. Both of these showers produce meteors at about a peak rate of 20 to 40 per hour.

1966 LEONID METEOR SHOWER. An incredible number of meteors are seen in this 12-minute exposure taken on November 17, 1966. Photograph by Scott Murrell, used with permission.

Depending upon the activity level and age of the comet, in some meteor showers the dust grains haven't had time to spread out over the entire orbit, but travel in large clumps that more or less travel around the sun with the parent comet. One spectacular example of this is the Leonid shower, which appears each year in mid-November and which is associated with Comet Tempel-Tuttle, a comet with an orbital period of 33 years. Normally, the Leonid shower is weak and unimpressive, with a peak rate of 10 meteors or less per hour, but around the time that Comet Tempel-Tuttle returns the strength of the shower peaks dramatically, producing on occasion not just a meteor "shower," but an out-and-out meteor "storm." One such Leonid storm occurred in 1833 when, for a brief period of time as observed from the eastern United States, up to a hundred thousand or more meteors per hour appeared; this was the famous night when "stars fell on Alabama." Another spectacular display occurred in 1866 – right after Comet Tempel-Tuttle was discovered – and although the displays around 1899 and 1932 were fairly poor, another very dramatic Leonid storm occurred in 1966, one year after the comet had passed perihelion. (The peak rate of almost 150,000 meteors per hour happened to occur over the western United States, and I remember this display well, thanks to an early-rising father who happened to notice the happenings in the sky and who came and roused me out of bed.) Comet Tempel-Tuttle is due to pass perihelion again in 1998, and there are hopes for yet another strong storm from the Leonids in either that or the following year.

COMETS AND ASTEROIDS, or, THE DEATH OF COMETS

As a comet makes its repeated trips around the sun, ejecting from its supply of gas and dust during each visit, eventually it gets to the point where its supplies run out. What happens then? There are still no good answers to this question, since comets have not been studied from a scientific standpoint long enough to derive valid statistics on just what happens when a comet dies. Perhaps the answer varies somewhat from comet to comet, depending upon the relative ratios of gas and dust within their respective nuclei; from the few cases of apparent cometary death that have been seen so far, there does appear to be more than one end result.

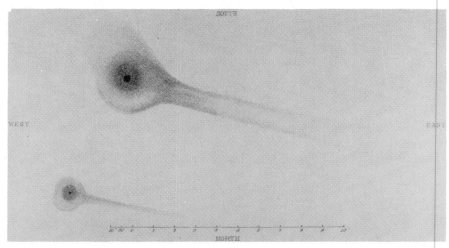

BIELA'S COMET. A sketch of this comet made in 1846, after it was found to be composed of two distinct components. Courtesy Don Yeomans.

Consider the rather spectacular case of Biela's Comet, a Jupiter-family object with a period of 6 2/3 years that was observed for several returns during the late 18th and early 19th Centuries. The comet startled everyone at its 1846 return by appearing not as one comet, but as two separate comets traveling through space together. The same thing was noted at the comet's next return in

1852, but after that... nothing; the comet was never seen again. However, there were intense meteor showers from the constellation Andromeda in 1872, 1885, 1892 and 1899 – all years during which Biela's Comet might be expected to return – and astronomers determined that the orbital characteristics of these meteors matched those of the comet. There seems to be little doubt, then, that Biela's Comet has completely disintegrated, and that these late-19th Century meteor showers were its final curtain call.

There has been a school of thought operating for about the past 30 years which states that, for at least some comets, as the nucleus gets baked with each passage around the sun, eventually it develops a hard "crust" which causes it to "shut off" completely. Such a "dormant" or "extinct" comet would appear, at least from a distance, as an object indistinguishable from an asteroid, and this particular idea has been used to explain the relatively high number of asteroids in the earth's vicinity, some of which do appear to travel in orbits rather reminiscent of comets.

This idea is somewhat hard to test observationally, but there do seem to be a couple of cases which suggest that there is at least some truth to it. Perhaps the best example is Comet Wilson-Harrington, a rather faint object which appeared on a few photographs taken over a period of seven nights in 1949. Although these were enough to show that the comet moved in a rather short-period orbit, the observations didn't quite cover a long enough period to tell just how short that period was, and this comet soon came to be regarded as "lost." Recently, however, some astronomers have determined that a faint asteroid discovered in 1979 and which moves in a somewhat comet-like orbit with a period of four years is in fact the same object as Comet Wilson-Harrington. Except for that one brief instant in 1949, when an obvious tail is present on the photographs, this object behaves as an otherwise "ordinary" asteroid, although the 1949 activity suggests that it does occasionally undergo episodes of cometary outgassing.

Then there is the case of the Geminid meteor shower, one of the year's most intense displays – with typical rates of 60 to 100 meteors per hour – that peaks each year during mid-December. Unlike the other strong showers, the Geminids didn't seem to have any particular comet associated with them. Then, in 1983, the *Infrared Astronomical Satellite (IRAS)* – which, incidentally, revolutionized many facets of astronomy during the 10 months it operated – detected a small asteroid, since named "Phaethon," that appeared to be traveling in the same orbit as the Geminids. This would seem to suggest that Phaethon is in fact an extinct or dormant comet, and after its discovery it was highly touted as such. A year later it passed close to the earth and was heavily scrutinized. The studies that were made then gave the somewhat surprising result that, as far as its surface characteristics and composition are concerned, Phaethon seems to be a typical "ordinary" asteroid rather than exhibiting any of the characteristics that an extinct comet might be expected to have. In fact, as one astronomer put it, except for its association with the Geminids, Phaethon doesn't seem to exhibit *any* cometary characteristics. If this is true, this leaves us with the question of how an asteroid can produce a meteor shower, particularly one as strong as the Geminids, and this question is still being pursued at this time.

COMETS AND THE EARTH

The idea that comets and asteroids can hit the earth, perhaps with catastrophic effects, is a relatively recent one, although it has been a staple of science fiction for some time. (For two of the better such stories that have been written, I recommend *Lucifer's Hammer* by Larry Niven and Jerry Pournelle, and *The Hammer of God* by Arthur C. Clarke. Information on both of these is given in Appendix A.) In a way, this is somewhat surprising, since all one has to do is check out our nearest neighbor, the moon and its menagarie of impact craters, to see that space in our part of the solar system is a dangerous place. The earth, with over thirteen times the surface area of our moon, should be hit more than the moon by about the same amount; it's just that, here on the earth, geological processes such as weathering (including wind and water erosion) and crustal movements wear down the impact craters after a (geologically) brief period of time. On the moon, where no such activity takes place, the impact craters remain pretty much in their original state for æons.

The realization that impacts by comets and asteroids can and do play a role in the earth's natural history has taken two parallel paths. The first one began about 1980 when the father-and-son geologist team of Luis and Walter Alvarez announced that rock sediments from 65 million years ago were heavily enriched with the element Iridium, which is quite rare on the earth but which is fairly common in meteorites. Now it so happens that one of the biggest mysteries in the earth's biological history concerns how the dinosaurs, who had ruled the earth for 130 million years, all died out relatively suddenly at about this same time. The Alvarezes hypothesized that a good-sized asteroid, say 6 to 10 miles across, collided with the earth 65 million years ago. As a result of the blast waves and earthquakes generated by this impact, as well as a global cooling caused by smoke from resulting forest fires and dust thrown into the atmosphere – a scenario not unlike those recently discussed under the term "nuclear winter" – most of the life forms on the earth, including the dinosaurs, died.

A theory as dramatic as this is sure to generate a lot of discussion and controversy among the scientific community, and this was no exception. By the present time a significant amount of evidence that supports this theory has been brought to light, and most, although certainly not all, scientists have now come to accept it as an event that probably happened. Perhaps the strongest evidence came in 1991 when a 110-mile-wide crater was identified lying underneath the surface rocks near the village of Puerto Chicxulub on the northern coast of the

THE MOON. Our nearest neighbor in space, showing the result of four billion years of asteroidal and cometary impacts. Copyright Lick Observatory.

Yucatán peninsula in Mexico. This is about the expected size of a crater that would result from the impact of an object in the size range that the Alvarezes discussed. Furthermore, rock samples taken from the crater indicate that it is precisely the same age as the Iridium-enriched rocks found in various other parts of the world. Many scientists now consider the Chicxulub crater as the "smoking gun" that provides final proof of the Alvarezes' hypothesis.

Besides the dinosaurs, other "mass extinctions" in the earth's history have been identified in the fossil record, and it is possible that these were also the results of impacts. Despite erosion processes, several impact craters have now been identified on the earth's surface; the most famous of these is the Meteor Crater (sometimes called Barringer Crater) near Winslow, Arizona, that appears to have been created by the impact of a 150-foot-wide asteroid about 49,000 years ago. Even today, we see occasional impacts; in June 1908 what was apparently a small asteroid exploded above a remote section of Siberia, flattening trees for hundreds of square miles around, and setting off blast waves

METEOR CRATER IN ARIZONA. A relatively fresh impact crater on the earth's surface. Courtesy Meteor Crater, Northern Arizona, USA.

in the atmosphere that circled the earth twice. The western United States may have escaped a similar fate in August 1972 when a similar-sized object entered the atmosphere, passed within 60 miles of the surface above Montana, then headed back out into space.

The second line of investigation has been astronomical, and began about the same time as the Alvarezes published their theory. Prior to about 1980 the discovery of an asteroid whose orbit brought it near the earth was a relatively rare event, and usually this happened no more than two or three times per year. Since that time, though, primarily as a result of dedicated searches such as those conducted by the Shoemakers at Palomar, the discovery rate of these objects has increased enormously, and today it is not at all unusual for three or four or more to be discovered during any given month. Some of these objects have indeed come quite close to the earth; the closest approach so far – at least, at the time of this writing – came in December 1994 when a tiny object, probably no more than 20 to 30 feet across, missed us by a scant 64,000 miles, barely over a quarter of the distance to the moon. (This object, incidentally, was discovered by the Spacewatch telescope, an automated search program that has been operating at Kitt Peak Observatory in Arizona since 1990.)

EARLY PHOTOGRAPH OF COMET SHOEMAKER-LEVY 9. This image was taken by Jim Scotti on March 30, 1993 with the Spacewatch telescope in Arizona. Used with permission.

Perhaps the most dramatic object that has been found during these various search programs was an object discovered by the Shoemakers during March 1993. This object, which Carolyn Shoemaker originally called a "squashed comet" when she first noticed it on the photographic films, was soon found to be a string of over twenty cometary nuclei, which had apparently resulted when the comet had passed close to Jupiter the previous year and had been ripped apart by that planet's immense gravity. At the time of its discovery Comet Shoemaker-Levy 9 – the Shoemakers were sometimes assisted by David Levy, and this was the ninth periodic comet found by this team – was in orbit around Jupiter, and was near the farthest point in that orbit (in other words, near "apojove"). In what was surely one of the most dramatic astronomical events that has ever been witnessed, each of the twenty or more fragments collided with Jupiter during the third week of July 1994, each impact liberating up to six million megatons of energy – several hundred thousand times greater than the largest man-made nuclear explosion – and leaving gigantic impact scars like black eyes in Jupiter's atmosphere that were plainly visible even in small backyard telescopes. It has been estimated that the largest fragment was probably no more than about half a mile in diameter, and it is a sobering thought that objects this small were able to leave scars that were quite a bit larger than the entire planet Earth. Many who watched these events unfold were struck by the realization that, the next time, it could be Earth that gets hit.

IMPACT! One of the fragments of Comet Shoemaker-Levy 9 collides with Jupiter in July 1994. Image taken by Peter McGregor, provided courtesy of the Mount Stromlo and Siding Spring Observatories.

Determining exactly what to do if an object is found to be on a collision course with the earth is a matter of much debate these days, and is beyond the scope of this book. One thing that can and is being done right now is the continuation of the search efforts described above, in order to obtain as complete a census as is possible of any potentially threatening objects.

Astronomers estimate that only about 10% of the asteroids (and comets) in the earth's vicinity have been discovered so far, but with the presently ongoing search programs, and others that are currently being planned, we can hope to have a reasonably complete census of these objects by the latter part of the 21st Century.

One potential future Earth collision has in fact already been identified. The culprit in this case is Comet Swift-Tuttle, the parent comet of the Perseid meteors. The comet's orbit and the earth's orbit intersect each other, suggesting that at some future date it is possible that both objects will arrive at the same point at the same time, resulting in a collision. After the comet was seen at its last return in 1992, there was even some speculation that such a collision was possible at its next return in 2126. More detailed studies of its orbit since then have shown that the comet will miss the earth by the rather comfortable margin of 14 million miles at that time. But in the meantime, an analysis of Comet Swift-Tuttle's future motion by astronomer John Chambers at the Center for Astrophysics in Cambridge, Massachusetts has revealed that, on or about September 15, 4479, the comet will pass so close to the earth that "...it is impossible to predict [the comet's] motion after that point" and that a collision is indeed possible at that time. Observations at future returns of the comet should help to clarify this one way or the other, but perhaps our descendants in that far-off era will feel some gratitude toward our generation for providing the advance warning.

SOME NOTABLE COMETS OF THE RECENT PAST, or, WHAT MAKES A GREAT COMET

Bright and spectacular comets have appeared on occasion throughout recorded history, and the records of the ancient and/or medieval Babylonians, Europeans, and – especially – the Chinese contain accounts of numerous appearances of these objects. Since many of these accounts are intertwined with the various mythologies of the observers, it is sometimes difficult to separate fact from fiction and to determine just what should be taken at face value. Nevertheless, there are some accounts of what truly appear to be immense comets that have made appearances throughout the centuries. The Chinese (and others) recorded comets in 1264 and 1402 which seem to rank among the best of all time; the former of these objects apparently was bright enough to be visible in daylight for some time, and had a tail which at one time spanned over 100° in length. (For comparison, the horizon-to-zenith distance is 90°.) Another spectacular object was de Cheseaux's Comet that appeared in 1744; at its brightest it outshone the planet Venus and developed a system of six distinct tails which spread out in the shape of a broad Oriental-style fan.

In discussing the potential that is offered by Comet Hale-Bopp it is instructive to examine some of the past comets to see what kinds of comparisons might be made. To be most useful, we should look at those characteristics that can cause the appearance of a comet to be spectacular (or, in some instances, those that cause a comet *not* to be), and see how some past comets with similar characteristics have performed. In many ways this type of comparative activity can be considered a "who's who" (or perhaps a "what's what") of the great comets of the recent past.

THE GREAT COMET OF 1744. Six tails of this object are seen extending up from the horizon prior to comet-rise. Courtesy Fred Whipple.

Not all comets come in the same size; just like with almost any other natural object, some are large, some are small, and many are in between. While it may not be exactly true in all cases, it is logical to suppose that the larger comets are also the brighter ones. In fact, astronomers don't usually speak too much in terms of the physical "size" of a comet, but rather they discuss its *intrinsic brightness;* i.e., its brightness in and of itself, without regard for any external factors. By way of analogy, think of those bright searchlights sometimes used by car dealerships for advertising purposes, and compare these to an ordinary flashlight; if all other things are equal, the searchlight far outshines the flashlight. So it is with comets.

Intrinsically, one of the brightest comets that has ever been seen was the Great Comet of 1811. This object never got too close to the sun (perihelion distance: 1.04 AU) nor to the earth (closest approach: 1.22 AU) and yet because of its immense size and intrinsic brightness it became as bright as the brightest stars in the sky. At one point the apparent coma was larger than the sun – this being the largest cometary coma ever observed – and at its greatest length the tail was 1.3 AU long, stretching over 70° of sky. The comet was so impressive to the general public that it received mention in much of the literature of the time, including Leo Tolstoy's novel *War and Peace*. (We'll look at this particular comet in a little more detail in Chapter 3.)

Another factor which can affect how bright a comet might get is its distance from the sun. The closer it gets, the more heat it receives, and thus its outgassing and dust eruptions proceed more vigorously. Also, the mere fact that it receives more light from the sun causes the dust in the coma to reflect more of this sunlight. Many of the "great" comets that have appeared in history were bright, not because they were intrinsically large objects, but because their perihelion distances were quite small.

During the past couple of centuries there have appeared several comets all traveling along the same basic orbit. They are not the same object – each has an orbital period of several centuries – but they are believed to have resulted from the breakup of some larger comet sometime in the distant past. This particular orbit brings the comets in extremely close to the sun – less than 0.01 AU – and as a result these comets are collectively referred to as "sungrazers." Not unexpectedly, this extreme closeness to the sun has caused several of the sungrazers to become truly spectacular objects. One of these, the Great Comet of 1843, could be seen with the naked eye during daylight when near perihelion, and afterward developed a tail 2 AU long. Another sungrazer, the Great Comet of 1882, apparently became even brighter than the comet of 1843 when near perihelion, and was also a spectacular object; so was the sungrazing Comet Ikeya-Seki that appeared in 1965. Most of these objects – the comet of 1882 being somewhat of an exception – as well as some of the lesser sungrazers that have appeared, were bright only for a brief period of time; as soon as they left the sun's vicinity they "quieted down," and once that happened the show was essentially over.

For obvious reasons, the closer a comet comes to the earth, the brighter it will appear to those of us on the earth looking at it. There have been several instances in history when otherwise dim comets have been able to put on a pretty good display simply because they were close to the earth. (To go back to the flashlight/searchlight analogy, the flashlight seen from a distance of, say, 5 feet will appear brighter than the searchlight from, say, 10 miles, even though the searchlight is much brighter intrinsically.) One recent example of this was Comet IRAS-Araki-Alcock, which passed 0.031 AU (2.9 million miles) from the earth in May 1983, one of the closest cometary approaches that has ever been recorded. Although intrinsically a rather dim object, because of its proximity to the earth it became as bright as the stars of the Big Dipper, and exhibited an apparent coma several times larger than the full moon. (It never developed much of a tail, and thus it appeared more or less as a large diffuse cloud in the night sky.) Its nearness to the earth caused its apparent motion to be quite rapid, and in fact within a few days it had traveled over halfway across the sky.

The geometry of a comet's appearance, i.e., its position in the sky with respect to the sun as seen from the earth, can also affect how "great" the comet becomes. Even an intrinsically bright comet with a small perihelion distance will not appear all that spectacular to us if during the time it is brightest we have to look in the same general direction as the sun in order to see it. In such a case, the comet will be seen against a bright twilight background, and thus there is a little contrast with which to observe the coma and the tail. That same comet, viewed against a dark nighttime sky, would, on the other hand, appear very spectacular. Geometry can also affect other facets of a comet's appearance, including the apparent length and direction of the tail, and very often it is the tail that determines the "greatness" of a comet.

COMET IRAS-ARAKI-ALCOCK IN 1983. Due to the comet's rapid motion, the stars appear as trails. Photograph taken by Alan Gorski, used with permission.

Many in the general public were disappointed by the appearance of Halley's Comet in 1986; this disappointment may have been fed by accounts from grandparents, etc., who saw the comet in 1910. Part of this disappointment may be due to reasons having nothing to do with the comet itself – for example, our grandparents in 1910 didn't have to contend with as much "light pollution" from city lights, etc., as we have to contend with today – but much of it was also due to the relative geometry between the two returns. In 1910 the comet was still fairly easily visible when at perihelion, and when closest to the earth a month later it was sunward of us but because of where it was placed its tail swept from well in front of the earth to well past it. This perspective briefly caused the tail to appear enormously long in our sky; in fact, the *apparent* length (as opposed to the true physical length) of Halley's tail on that occasion was one of the longest that has ever been recorded for any comet. In 1986, by contrast, when Halley was at perihelion it was on the opposite side of the sun, and invisible. When it was closest to the earth two months later it was already beyond the earth's orbit, and from our vantage point the bulk of the tail extended directly behind the comet and was hidden by the coma; this created the "fuzzball" appearance that many readers might remember. In terms of its actual brightness, the comet really wasn't much brighter in 1910 than it was in 1986; it was the much better geometry that prevailed in 1910 that caused it to be so much more spectacular then. The geometry in 1986, in fact, was just about the worst possible for Halley's Comet; just our generation's bad luck!

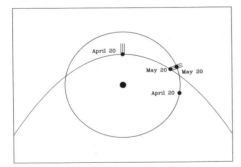 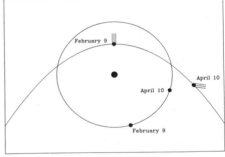

THE EFFECT OF GEOMETRY. The geometrical circumstances of Halley's Comet during its 1910 (left) and 1986 (right) returns. The relative geometry allowed the comet to be much more prominent in 1910 than it was in 1986.

The physical and chemical processes going on within a cometary nucleus can affect how bright the comet eventually becomes. Many readers will remember the disappointing display put on by Comet Kohoutek in 1974. When discovered, this object was almost 5 AU from the sun and appeared relatively bright for being such a far distance away, and its unusually small perihelion distance of 0.14 AU suggested that it would become a truly great object when it came near the sun. We all know that this did not happen; Comet Kohoutek did become visible to the naked eye (to someone who knew where to look and who had a dark sky available) but was nowhere near the "great comet" that many had expected it to be. Another recent "fizzle" was Comet Austin in 1990; like Kohoutek, it was relatively bright when discovered relatively far from the sun, and was traveling in an orbit which suggested that it would become quite bright. Comet Austin performed even worse than Comet Kohoutek did, becoming only barely visible to the naked eye even in a dark sky.

HALLEY'S COMET IN 1910. Venus is the prominent object to the comet's lower right. Lowell Observatory photograph.

As it turns out, both Comets Kohoutek and Austin were making their first visit sunward from the Oort Cloud. From studying these and other first-time Oort Cloud comets, we've learned that, during the 4½ billion years that the

COMET KOHOUTEK. This photograph taken in January 1974 shows the comet at about its best (as far as Earthbound viewers were concerned). Photo taken by Dennis di Cicco, used with permission.

comet sits in the cosmic deep-freeze near interstellar space, it collects a thick "crust" of interstellar dust and other materials as the sun's travels carry it around the galaxy. A small proportion of the icy molecules that outgas into a cometary coma will collect onto this crust after it develops. Then, as the comet travels sunward and starts to experience solar heating for the first time, this outer layer of ices will start to vaporize when still relatively far away. If the comet should be discovered at this time, as both Kohoutek and Austin were, we would see what appears to be an intrinsically bright comet located rather far from the sun. This brightening is only superficial, however, and once this outer layer of ices burns off, what we're left with is the nucleus covered by the thick crust, which more or less causes the comet to "shut down." If the comet should come reasonably close to the sun, this crust will get baked off as it makes its passage through the inner solar system, but unfortunately for those of us who were hoping for a brilliant cometary display, this is too little, too late. After the comet has been "broken in" by this first perihelion passage, it should perform well at future returns, however. If it's any consolation to those who were so disappointed by Comet Kohoutek in 1974, this object should put on an absolutely spectacular display when it makes its next return 75,000 years from now!

Sometimes the physical processes within a comet's nucleus can influence its brightness in a positive direction as well. If, for some reason, the nucleus should split into two or more separate pieces, a lot of previously-hidden ice and dust is exposed to sunlight for the first time. This will usually cause a tremendous increase in the amount of material being ejected, and most of the time this will cause a comet to flare up in brightness quite dramatically. Although such splitting can occur almost anywhere in a comet's orbit – it has been observed to happen relatively far from the sun – usually this happens close to the sun where there's a lot of solar heating "frying" the nucleus. This, in fact, is exactly what happened to Comet West in 1976 and caused it to be so spectacular. While it wasn't all that bright going in to perihelion – and consequently no one was expecting a brilliant display afterward – the intense solar heat at its perihelion distance of 0.2 AU caused the comet's nucleus to split into four separate pieces. This newly-exposed cometary surface created a tremendous upsurge in brightness, and this is what caused the spectacular morning-sky spectacle that many of us remember from that March.

THE BREAKUP OF COMET WEST. This series of photographs shows the nucleus of Comet West splitting up into four separate pieces, and subsequently spreading apart. These images were taken at New Mexico State University's Tortugas Mountain Observatory, and are provided courtesy Reta Beebe.

All of these various factors interact to produce the cometary display we see in our skies. A classic example of this occurred just as I was in the finishing stages of writing this book with the arrival of Comet Hyakutake (C/1996 B2), which many had the opportunity to view when it made its appearance in the spring of 1996. Several circumstances – including a moderately high intrinsic brightness, a small perihelion distance, a close approach to the earth, and excellent viewing geometry – combined to make this object one of the best comets that has appeared during the past few decades. Several of the lessons we observed in Comet Hyakutake's display can be applied to Comet Hale-Bopp when we start making predictions about the display it will produce, and for this reason I will say quite a bit more about Comet Hyakutake in Chapter 3.

Many of the different variables I've discussed can be determined ahead of time once a comet's orbit is computed; for example, we'll know how close a comet will get to the sun and to the earth, and we can calculate the effects due to geometry. We can also tell, after a month or two of obtaining positions for calculating an orbit, whether or not the comet is making its first trip from the Oort Cloud. But even here, our powers of prediction start to fail a little; there have been some comets fresh from the Oort Cloud that still have managed to perform rather well. There is, of course, no way to determine ahead of time whether or not a comet will split into pieces. What this all adds up to is that the behavior of a comet as it passes perihelion – especially a newly-discovered comet that hasn't been seen before – cannot be predicted with any reasonable accuracy. Fred Whipple, of the "dirty snowball" fame, once remarked that "if you must bet, bet on a horse, not on a comet," and there is a lot of truth in this. As far as Comet Hale-Bopp is concerned, about all we can do is calculate its orbit – which has been done – and, by using the comets I've talked about here, as well as others, as guides, make some *preliminary* estimates as to how bright and spectacular it might eventually become. But, at this stage, almost anything can happen, and as I've been saying to the people who discuss this with me, "the comet's going to do whatever the comet's going to do." For now, we have to leave it at that.

CHAPTER 2: THE EARLY HISTORY OF COMET HALE-BOPP

The most exciting phrase to hear in science, the one that heralds new discoveries, is not "Eureka!" (I found it!) but "That's funny..."
  – Isaac Asimov*

THE DISCOVERY BY HALE

As I mentioned in Chapter 1, on any given night two or three comets are visible, on the average, in a decently-sized backyard telescope. Most of these are dim and inconspicuous objects, and of no real interest except to those scientists who study comets. But I happen to be one of those; I began observing comets in 1970, and for the past 20 years I have been making brightness measurements of them and forwarding these on to the International Astronomical Union's (IAU's) Central Bureau for Astronomical Telegrams in Cambridge, Massachusetts, who then publish them in various publications for usage by professional cometary scientists in their research. Even though I have earned a Ph.D., and work professionally, in a different branch of astronomy, I still continue to observe comets as a major pastime.

July and August usually make up the rainy season – sometimes called the "monsoon" season – in New Mexico. Due to the typical wind flow at that time of the year, moist air from the Gulf of Mexico blows northwestward until it meets the southern Rocky Mountains, at which point it rises and allows the moisture to condense out. This usually produces thunderstorms – which are sometimes very intense – throughout the southwestern part of the United States; these storms often occur on a daily basis. As one might imagine, clear nights are rather rare while all this is going on.

Saturday night – the night of July 22-23, 1995 – was the first clear night we had experienced here in Cloudcroft, New Mexico, for about a week and a half, and was the first clear night since the full moon. (Astronomers usually don't observe much while the moon is full, since the resulting bright sky wipes out the dimmer objects we usually like to study. Most of the time, this is when we catch up on our sleep.) My normal practice is to observe any comets I might be following about once a week, and since I had no idea as to when I might get another clear night, I decided to take advantage of the clear weather in order to collect observations of the two comets I was then following. Both of these were rather dim objects, although one of them was expected to become bright enough to be visible in binoculars in a few weeks, and I was planning to write about this

---

* Isaac Asimov is one of my favorite writers, and his writings profoundly influenced my way of viewing the world. Interestingly, I came across this quote only a couple of weeks before my discovery of Comet Hale-Bopp.

object in the next week's edition of the weekly astronomy column I write for one of the local newspapers.

THE DISCOVERY SITE. The scene from my driveway, where I discovered Comet Hale-Bopp with the telescope shown.

Having just moved into our family's new house outside of Cloudcroft a few months earlier, I had not yet built any kind of observatory. Instead, I keep my telescope in the garage and wheel it out onto the driveway whenever I want to use it. Thus, a little after 11:00 PM that evening I did just that, and commenced to observe the two comets. I located the first one and, as I always do during such an observation, measured its brightness. I finished with it just before midnight.

Even though the second comet had already risen by this time, it was still somewhat low in the southeast. From the driveway where I had the telescope set up, our house blocks the view in that direction, and I still had approximately an hour to an hour and a half before the comet would rise high enough to look at. I thus had some time to kill, and I was tempted to go indoors and watch television or read. But as it was the night was of the beautiful, crystal-clear variety that provided part of our rationale for moving up into the southern New Mexico mountains. The Milky Way, in particular, was especially impressive as it swept from the north, through the point overhead, and then on into the south, where its brightest section was marked by the prominent constellation Sagittarius. Within Sagittarius there are quite a few star clusters and other so-called "deep-sky objects," many of which can be quite impressive when viewed through a decent telescope. I had looked at some of these objects one evening a couple of weeks earlier, and I thought that I would pass the time by re-examining some of them.

I turned the telescope to one such star cluster, which is catalogued under the name "M70," and immediately saw a dimmer, fuzzy object in the same field of view. I had looked at M70 during that evening two weeks before, and I didn't recall seeing such an object near it at that time. Not being sure what to make of it, the first thing I did was examine the object with higher magnification; it can happen that a group of stars close together can appear fuzzy when viewed at a low power. But, the object still appeared fuzzy at the higher power, and didn't show any sign of being made up of stars. I then thought that, instead of looking at M70, I might have inadvertently pointed the telescope toward a different star cluster (of which there exists a profusion in that part of Sagittarius), but a check of my star atlas revealed that I was indeed looking at M70, and that there weren't any other objects plotted near it.

By this time I was strongly suspecting a comet, despite the fact that finding a comet in this manner is an extremely unlikely proposition. After all, many amateur astronomers scan the skies for comets on a nightly basis, and may go years, if not a lifetime, and not find anything; I myself had spent several years in this pursuit, without success. Also, this object was in a part of the sky that one might expect to be covered in the course of an observatory's search program, such as that conducted by Eugene Shoemaker (although this particular program had been terminated at the end of 1994, there are others going on). Still, stranger things have happened, and past experience has taught me to check out any suspicious object I might come across during my observations.

The acid test as to whether or not this would be a comet is motion; if indeed it was a comet, it would move against the background stars over a period of time (say, an hour or two). Thus, I made a pencil sketch showing the object's position relative to the stars that were close by in the field. And, since I was already observing other comets that night and measuring their brightnesses, I went ahead and measured this object's brightness as well.

By the time I had finished all this, about half an hour had elapsed since my first sighting of it. I actually thought I might be seeing motion by this time, but since I was truly hoping this was a comet, I recognized this might be nothing more than wishful thinking.

At that time I went indoors and upstairs to my home office, where I have a relatively extensive astronomical library. Although my star atlas hadn't shown anything plotted in the position of my object, that didn't necessarily mean anything; there are plenty of such objects which, for some reason or other (usually dimness), aren't plotted in most star atlases. (It in fact would be quite impractical to do so.) Among the books in my library are several catalogues of deep-sky objects (star clusters, galaxies, etc.), and I spent several minutes going through these to see if my object might be listed in any of them. I found no listing of it anywhere.

THE DISCOVERY POSITION OF COMET HALE-BOPP. The star cluster M70 is in the upper left, and the "X" marks the location where I first noticed the fuzzy object that wasn't supposed to be there. Copyright Anglo-Australian Observatory/Royal Observatory, Edinburgh, used with permission. This image was obtained from the Digitized Sky Survey, produced at the Space Telescope Science Institute under U.S. Government grant NAGW-2166, and processed into the present compressed digital form with the permission of the AAO.

I was quite convinced by this time that it was indeed a comet, but the possibility still remained that it was an already-known one. Although I try to keep track of any comets I might expect to be visible in my telescope, it is possible that one of the extremely dim comets – of which there are always several in the sky – might have experienced a "flare," causing a temporary dramatic increase in its brightness. (Such flares, or "outbursts" as they are sometimes called, do occur in comets from time to time.) The main computer at the IAU Central Bureau's office in Cambridge contains a program wherein one can input a position on the sky (in the form of coordinates that are similar to latitude and longitude on the earth's surface) and it will return with any known comets that are in that position. I used my office computer to access the Internet, from which I then logged onto the IAU's computer to run their program. The program came back negative; no known comets were located at the position of my object. I also checked to see if there might have been a comet just discovered – the Central Bureau issues a publication called the IAU *Circulars* which, among other things, announce the discoveries of new comets – and this also turned up negative.

COMET HALE-BOPP AND M70. *The comet is the small fuzzy object close to the center. This photograph was taken by Kesao Takamizawa only a few hours after the comet's discovery by Tom Bopp and myself. Used with permission.*

Since I still hadn't definitely seen motion yet, I wasn't 100% sure I had a new comet, but I felt by this time that my chances were pretty good. So, before I logged off my computer I sent an email message to Brian Marsden and Dan

Green – the Director and Assistant Director, respectively, of the Central Bureau – informing them of my discovery of a possible comet. In my message I gave them an approximate position for the object, as well as an approximate brightness; I also stated that I thought I might have seen motion, but would report back later when I had verified this one way or the other.

Once I finished this and logged off, I went back downstairs and outside. When I looked through the telescope at my object I saw that it indeed had moved slightly to the west – no doubt about it. I immediately went back inside, logged on to my computer again, and sent an email to that effect to Marsden and Green.

Sure by now that I had a new comet, and having now reported it, I embarked on a risky endeavor: I went into my bedroom, woke up my sleeping wife Eva, and asked her if she was interested in looking at Comet Hale. I'm not quite sure if she believed me or not, but she did come down to the telescope to look at the object. Her sister Alice happened to be visiting us at the time, and she, too, came out to look at the comet after Eva woke her up. On the other hand, when Eva woke up our older son, Zachary (then 8 years old), he muttered something about "Dad and his spacey stuff again" and went back to sleep.

The women went back to bed after a few minutes, but I continued to stay up and watch the comet as it continued its slow motion against the background stars. (Incidentally, I did take some time out to observe the other comet that I had originally been waiting for.) Finally, a little after 3:00 AM, the comet set behind the trees in the southwest. Since I now had a three-hour baseline of positions for the comet, I thought I had enough information to file a detailed report, but before I went inside I took a moment to stand quietly under the stars, open my arms slightly, and utter a quiet "thank you" to the sky for providing this comet to me.

After going indoors, I used my star atlases to carefully measure the positions of the comet for the time I had first seen it, and for just before it had set. I then sent a detailed report of these via a third email to Marsden and Green at the Central Bureau. Afterward, I made an attempt to grab a few hours of sleep, but my state of excitement was so high that I was unable to do more than take an occasional doze.

Finally, at around 8:00 that morning I got out of bed, strolled over to the computer, and logged onto my Internet account. There was an email message from Brian Marsden stating "Congratulations! Did you ask Thomas Boppe [*sic*] in Glendale to confirm it? He seems to have done so." I replied back stating that I didn't know a Thomas Boppe in Glendale, and then logged off. (To be sure, I didn't even know which Glendale Marsden might be referring to; I thought it was probably Glendale, California.) Shortly thereafter, Eva, our two sons, Alice, and I left for Cloudcroft to have Sunday brunch at one of the local restaurants and to go hiking on one of the mountain trails.

We returned to our house around 2:00 that afternoon, and I immediately checked my email. Sure enough, there was a congratulatory message from Dan Green, which gave me a few details about Thomas Bopp's discovery. And then, the moment I was waiting for: Green had just issued IAU *Circular* 6187, which announced our discoveries of the comet to the astronomical world.

Circular No. 6187

Central Bureau for Astronomical Telegrams
INTERNATIONAL ASTRONOMICAL UNION

Postal Address: Central Bureau for Astronomical Telegrams
Smithsonian Astrophysical Observatory, Cambridge, MA 02138, U.S.A.
IAUSUBS@CFA.HARVARD.EDU or FAX 617-495-7231 (subscriptions)
BMARSDEN@CFA.HARVARD.EDU or DGREEN@CFA.HARVARD.EDU (science)
Phone 617-495-7244/7440/7444 (for emergency use only)

COMET 1995 O1

Independent reports of the visual discoveries of a new comet have been received from Alan Hale and Thomas Bopp, with available observations given below. All observers note the comet to be diffuse with some condensation and no tail, motion toward the west-northwest.

1995 UT	α_{2000}	δ_{2000}	m_1	Observer
July 23.264	18^h44^m23	$-32°13.6$	10.5	Hale
23.30	18 46	-32.6	10.8	Stevens
23.375	18 44.17	$-32\ 11.2$		Hale

A. Hale (Cloudcroft, NM). 0.41-m reflector.
J. Stevens and T. Bopp (near Stanfield, AZ). 0.44-m $f/4.5$ Dobsonian reflector. Comet found while observing M70.

PSR J0538+2817

X. Sun, Max-Planck-Institut für Extraterrestrische Physik (MPIEP) and Institute of High Energy Physics, Beijing; B. Aschenbach, W. Becker, and J. Trümper, MPIEP; and S. Anderson and A. Wolszczan, Pennsylvania State University, report the discovery of x-ray emission from the radio pulsar PSR J0538+2817 in S147 (= G180.0−1.7; cf. *IAUC* 6012) at count rates of 0.048 ± 0.013 and 0.051 ± 0.012 count/s in the broad (0.8–2.0 keV) and hard (0.5–2.0 keV) energy bands, respectively: "The x-ray source was detected during the ROSAT All Sky Survey at $\alpha = 5^h38^m26.0$, $\delta = +28°17'15''$ (equinox 2000.0), which is only 14'' away from the radio pulsar (Anderson *et al.* 1995, work in progress), well within the error of the ROSAT PSPC position. There is no other object in the vicinity of the source in the SIMBAD and GSC databases. The short survey exposure time does not permit a timing analysis or a search for a compact nebula emission, although a large complex diffuse emission associated with SNR S147 was detected in the hard-energy band. Assuming a blackbody spectrum and a 10-km-radius emission area, and adopting the column density (1.2×10^{21}) and distance (1.5 kpc) derived from the radio-dispersion measure, the energy distribution of photons can be used to constrain the temperature and luminosity to ~ 0.1 keV and 3×10^{32} erg/s. This temperature is in rough agreement with the prediction of standard neutron star cooling models using the spin-down age of 6×10^5 yr."

1995 July 23   *Daniel W. E. Green*

IAU Circular *6187, announcing the discovery of Comet Hale-Bopp. Copyright Central Bureau for Astronomical Telegrams, used with permission.*

THE DISCOVERY BY BOPP

Unlike me, Tom Bopp had never observed a comet before that fateful night in July. However, he has been an amateur astronomer for over 25 years, and like many others, he takes a special enjoyment in looking at the various galaxies, star clusters, and other deep-sky objects the sky has to offer. He relocated to the Phoenix area from his native Ohio several years ago, and unfortunately from there he is not able to enjoy the night sky as well as he would like; the "light pollution" from all the streetlights, billboards, etc. that are endemic to a major metropolitan area like Phoenix keeps the night sky from becoming truly "dark." Thus, like many other amateur astronomers whose work forces them to live within large urban areas, when he wants to view objects in the night sky he must drive out a good distance into the "sticks" to get away from all the city's light.

On that particular Saturday night Tom and some of his friends decided to take advantage of the break from the summer monsoons – which hit Arizona just like they hit New Mexico – and (in separate cars) they drove to a spot in the Arizona desert some 90 miles south of Phoenix. Several of the group brought telescopes, although Tom, who at that particular time did not own one, did not. Individually, and as a group, the various folks who were there examined several of the deep-sky objects that were visible at the time, until finally Jim Stevens, who owned one of the larger telescopes set up there – a telescope he had built himself, in fact – decided to examine some of the star clusters in Sagittarius. He and Tom took turns looking at the various clusters, until finally, right around 11:00 PM, they came to M70. (Keep in mind that, while New Mexico "springs forward" to daylight savings time, Arizona doesn't. These events were thus occurring at the same time as my goings-on a few hundred miles away.)

After Stevens had looked at the cluster, he began examining his star atlas to find his next "target;" meanwhile, Tom was taking his turn at the eyepiece. Because of the earth's rotation, any object a telescope is pointed at will move out of the field of view within a few minutes unless the telescope is electrically driven to follow that rotation (or unless the person at the telescope moves it manually). Stevens' telescope didn't have the necessary electrical equipment, and consequently M70 began to move out of the field of view as Tom was watching it. Instead of moving the telescope by hand, he decided to just watch the cluster move out, and as he did so, he noticed another fuzzy object, somewhat dimmer than M70, coming into the field of view.

At that point Tom asked his friend if there were any other objects plotted near M70. When Stevens replied that there weren't, Tom began to suspect that he might have a comet, especially after examining the object at higher magnification and verifying that it wasn't just a group of dim stars close together. By this time several of the others in the group became aware of the animated conversation going on over at Stevens' telescope, and after hearing about the possible comet discovery several of them were able to examine the object in their own telescopes and verify its existence. Tom has told me that some in the group thought that they should just dismiss it – it was almost certainly a deep-sky object that just didn't get plotted in that particular star atlas – but he decided that it was worth checking for motion.

For the next hour, Tom, Stevens, and a couple of the others there watched the object, and after that time they were convinced they had seen it move slightly to the west. There then remained the matter of reporting it, but out in the desert – almost literally, in the middle of nowhere – this wasn't an easy task. Tom tried to call with his cellular phone, but was unable to get through to anywhere from their location. Finally, he got into his car and headed back toward Phoenix, with the intent of sending a telegram to the Central Bureau once he got home.

HALE AND BOPP. Thomas Bopp (left) and me, shortly after our first meeting which took place at the Enchanted Skies Star Party in Socorro, New Mexico on September 22, 1995. Photograph by Electra Field, provided courtesy of the Enchanted Skies Star Party and Dave Finley.

Along the way, he found a truck stop and stopped, thinking he might be able to send a telegram from a pay phone. Unfortunately, the representative at Western Union didn't have an address for the Central Bureau, and since Tom didn't have that with him, he had little choice but to hang up and resume the journey home. Once he arrived at his home in Glendale (just outside of Phoenix) he rummaged through his library until he found the correct address for the Central Bureau. He was then able to give this to Western Union, and could finally send a telegram to the Bureau informing them of his discovery of the comet.

Like me, Tom was unable to get much decent sleep after that. Later that morning Dan Green called him and asked him a few questions about the discovery; among other things, the Central Bureau wanted to verify that Tom and I hadn't collaborated in any way. After speaking on the phone with several of the others who were with Tom that night, Green was finally able to tell Tom what he wanted to hear: "Congratulations! I think you've discovered a new comet!"

THE DETERMINATION OF THE ORBIT

The discovery of a comet is only half the fun; the rest comes once the comet's orbit has been determined. This is what tells us, at least approximately, how bright the comet will get, how long we (the world's astronomers, that is) will be able to follow it, and so on. My first thought, once I had firmly recognized the new object as a comet and had seen its motion, was that this was quite possibly a new member of Jupiter's family; most such objects are on low-inclination orbits which keep them near the *ecliptic;* i.e., the sun's path across the sky. (The planets stick close to the ecliptic during their travels as well, and the particular group of constellations that the planets usually occupy is often called the *zodiac*.) Sagittarius, as many readers may recognize, is a constellation of the zodiac, and since the new comet's motion was roughly parallel with the ecliptic, it seemed rather likely that I was indeed looking at a new short-period comet ("new" in the sense that it hadn't been seen before, although it was also possible that this might be a comet that had recently been "captured" into a short-period orbit). As I mentioned in Chapter 1, discoveries of such objects are made every year, and while my new comet was somewhat brighter than the typical such discovery, a Jupiter-family comet of this brightness is not unprecedented. What this probably would mean was that my comet wouldn't get any brighter than it was at the time, and, except for the scientists who are specifically interested in comets (and possibly the local news media, for perhaps a day or two), nobody would pay much much attention to this discovery. On the other hand, this also meant that, every few years, comet scientists would be confronted with the name "Hale" whenever the object returned to perihelion, and I decided I could probably live with that.

Once a comet is discovered, astronomers at several locations around the world will take images of it in order to measure its position as precisely as possible; once enough of these measurements are obtained, the comet's orbit

COMET HALE-BOPP, taken on July 24, 1995 – the night after my discovery – by Warren Offutt of W&B Observatory in Cloudcroft, New Mexico. This was one of the earliest images used in the calculation of Hale-Bopp's orbit. Image courtesy Warren Offutt.

can be computed. Usually this takes only two or three days, at which time the orbit is announced on the IAU *Circulars* and the comet is named. (When the IAU introduced its new designation system at the beginning of 1995, it also initiated the policy of not naming a new comet until the initial orbit is available. Among other reasons, it sometimes happens that a newly-discovered comet turns out to be a long-lost periodic object, and the IAU wants to avoid adding names to a comet unnecessarily.)

The first indication that there might be something unusual about Comet Hale-Bopp came in an email message Brian Marsden sent me on Monday, July 24, wherein he commented that there was no discernible parallax in near-simultaneous observations made from Australia and Japan. Recall in Chapter 1 that it was Tycho Brahe's inability to detect a parallax in his observations of a bright comet in 1577 which showed that comets are phenomena of the solar system, not of the earth's atmosphere. This is true enough in the precision that can be obtained in observations made with the naked eye, but with the precise measurements that can be obtained with good telescope equipment today it is usually possible to measure a parallax for objects which aren't too far from the earth. The fact that it wasn't possible to do so in the case of Comet Hale-Bopp indicated it was a respectable distance away; Brian suggested that it was at least 2 AU from the earth at that time. Among other things, this suggested that the comet was unlikely to be a periodic object in Jupiter's family; such comets are usually quite dim intrinsically, and would not be bright enough to be visible in a small telescope from that far away.

Even by a couple of days later the comet's orbit was still almost complete *indeterminate;* i.e., almost any conceivable type of orbit, from a circle to a hyperbola, could fit the positions received so far. To illustrate this, on Tuesday afternoon Dan Green emailed me *ephemerides* (plural of *ephemeris*) resulting from two completely different orbits, one a parabola, the other a near-circle; even after several days the positions resulting from these two dramatically different orbits were almost indistinguishable.

Finally, by about mid-day on Wednesday (July 26), one type of orbit began to separate itself from among the host of orbits that were possible, but what this suggested seemed absolutely incredible: the comet apparently was well beyond the orbit of Jupiter (6.7 AU, as opposed to Jupiter's 5.2 AU), suggesting an incredible intrinsic brightness – intrinsically, just about the brightest comet that's ever been seen. Moreover, the comet wouldn't pass its perihelion point until early 1997, when it would be well inside the orbit of the earth (0.8 AU). If one took the comet's present brightness and extrapolated along this orbit, one was left with the possibility that, around the time of perihelion, Comet Hale-Bopp could well be brighter than the brightest stars in the sky – possibly as bright as the planet Venus – and that, not only could the comet become one of the brightest of the 20th Century, but perhaps one of the brightest in all of recorded history.

With this announcement the comet was officially named Comet Hale-Bopp, but when I saw the IAU *Circular* in question I didn't pay much attention to that; I noticed the orbit, which left me almost in a state of absolute shock. I had been hoping all along – as I'm sure most discoverers of new comets do – that my find would become a bright naked-eye spectacle, but something like this was clearly beyond my wildest fantasies. Still, I couldn't get too excited just yet; for one thing, there were still the infamous cases of Comets Kohoutek and Austin to think about (see Chapter 1), both of which had been relatively bright when relatively far from the sun, but had "fizzled" dramatically when they approached perihelion. (Both these objects were fainter, and were closer to the sun at their respective discoveries, though, than was Comet Hale-Bopp.) Even more important was the fact that the orbit was still extremely uncertain at that time; it was quite possible that the "true" orbit was significantly different from, and would indicate a cometary display dramatically less spectacular than, this initial very preliminary calculation.

Indeed, as more observations continued to pour in over the next few days, it began to look like the perihelion distance was quite a bit larger than that given by the initial orbit, indicating a much less spectacular object, quite possibly never becoming visible to the naked eye. In fact, according to an email message that Dan Green sent me on Thursday, it looked like the perihelion distance could be anywhere between 1 and 10 AU. After a couple of more days, that range seemed to be narrowing down to somewhere between 2 and 8 AU, and by the end of the week it began to appear as though perihelion would be in the range of 2 to 3 AU. While this type of an orbit might bring the comet up to naked-eye visibility, it would not be an especially bright object, and certainly nothing like the Great Comet of 1997 that the first orbit had suggested. My own feeling was something like what one experiences after the typical summer romance: it was nice while it lasted, but...

And then on Monday, July 31, I received an email from Brian Marsden telling me that the perihelion distance was starting to look like 1 AU again! On the following day, he issued IAU *Circular* 6194 where, from 208 separate positions obtained during the preceding nine days, there was an orbit indicating perihelion on April 1, 1997, at just slightly over 0.9 AU from the sun. Another interesting item about the orbit was that its inclination was almost exactly 90°; in other words, its orbit was almost exactly perpendicular to the earth's.

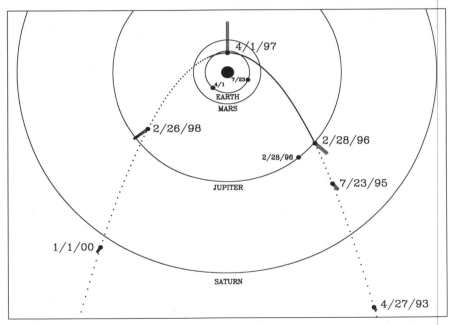

ORBIT OF COMET HALE-BOPP. The comet's orbit is shown, together with some of the major planets. Some important dates are indicated. This is a two-dimensional representation of a three-dimensional orbit; those sections of the comet's orbit which are north of the plane of the earth's orbit are indicated by a solid line, those sections south of the plane are indicated by dots.

Although this new orbit was still somewhat uncertain, there was some justification for putting faith in it, since it covered a much larger set of positions, over a longer period of time, than did the initial orbit. While the new orbit suggested a scenario in 1997 that wasn't quite as good as that the first orbit suggested, it nevertheless indicated a potentially bright display; the comet was at an almost-unprecedented 7.15 AU from the sun on the night that Tom Bopp and I discovered it, and while it wasn't going to be coming anywhere near the earth, its distance from the sun at perihelion suggested that it could easily become as bright, or brighter, than the brightest stars in the sky. As I put it at the time, it had gone from possibly being one of the best comets of all time to being one of the best of the century, and I decided that I could indeed live with that.

There remained the question of whether or not Comet Hale-Bopp was a fresh comet making its first visit in from the Oort Cloud; Comets Kohoutek and Austin were still pretty fresh in our minds at the time. Usually, though, this takes a while to determine; it normally requires a series of observations covering a span of several weeks. On the other hand, this process can be speeded up a bit if any *pre-discovery* images of the comet are found. These are photographs taken before the comet's discovery which in fact contain images of it, but for some reason or other the images were not noticed beforehand. It's a pretty standard practice among the world's comet astronomers to examine old photographs for such images once a valid orbit for a comet has been published; in several cases this process has been successful in extending the available positions of a comet backward in time several weeks or more.

After getting up on Wednesday morning, August 2, I checked my email, and there was a message from Rob McNaught at Siding Spring Observatory in New South Wales, Australia. (Rob is considered by many astronomers who study comets, myself included, to be one of the top observational astronomers in the world.) In this message – which was actually a "copy to" version of a message he had sent to Brian Marsden and Dan Green – Rob stated he had found a possible image of Comet Hale-Bopp on a photograph taken with one of the telescopes at Siding Spring as far back as April 27, 1993. At that time the comet was 13.1 AU from the sun – over 3 AU beyond the orbit of Saturn – and although its image was quite dim on the photograph – far too faint an object to have been visible in my own telescope – it was unprecedently bright for a comet at that distance. While there did exist the possibility that this image was just a blemish on the photographic emulsion, Rob – who is an expert at distinguishing such blemishes from "real" objects – did not seem to think this was the case. If this was indeed an image of Comet Hale-Bopp, it would extend the span of observations backwards in time over two years, and would thus allow a much more precise calculation of the comet's orbit.

Intrigued by this message, I sent Brian Marsden an email asking him his thoughts. He replied a while later that Rob's position did fit with the rest of the observations – in his words, "beautifully" – and, what was more important, indicated that Comet Hale-Bopp was *not* a fresh comet in from the Oort Cloud; it's been around at least once before. Later that day Brian issued another IAU *Circular* giving details about Rob's position, as well as the orbit which resulted when it was included in the calculations. There was actually little change in the perihelion date and distance, but the calculation suggested that the comet made its last appearance in the inner solar system a few thousand years ago.

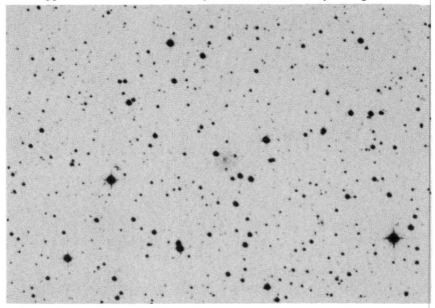

COMET HALE-BOPP IN 1993. This is the pre-discovery image taken on April 27, 1993, by Paul Cass at Siding Spring Observatory in New South Wales. (The image is a negative, showing dark stars on a white background). Comet Hale-Bopp is the small fuzzy object just above center. Copyright by the Anglo-Australian Observatory, image provided courtesy of Rob McNaught.

There still remained the one catch: we still couldn't be absolutely sure that Rob's 1993 position wasn't a blemish, and there was also the fact that the comet didn't seem to appear on a photograph taken at Siding Spring in September 1991, when it shouldn't have been too much dimmer than it was in 1993. Nevertheless, in the meantime observations continued to be reported to the Central Bureau, and by early September Brian was able to determine from the 1995 observations alone that Comet Hale-Bopp cannot be making its first visit in from the Oort Cloud. Finally, by late October enough observations had been obtained to show conclusively that the 1993 image is indeed of the comet. It seems, then, that we might have cause for optimism after all; at least, a "fizzle" like those by Comets Kohoutek and Austin seems to be less likely.

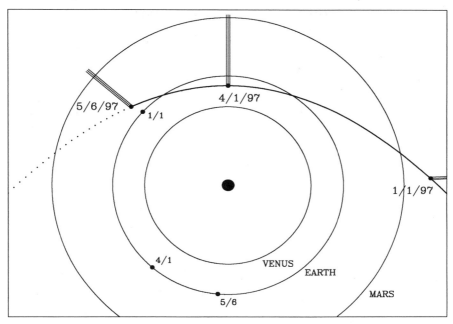

INNER PORTION OF HALE-BOPP'S ORBIT. A close-up of the diagram shown on page 60.

The most recent orbital calculation that's been published as I write this utilizes observations obtained up through late March 1996, and indicates that Comet Hale-Bopp will pass perihelion on April 1, 1997, at 3:30 Universal Time (sometimes called Greenwich Time, the time system usually used by astronomers for consistency); this corresponds to 8:30 PM Mountain Standard Time on March 31. (Expect this to change slightly – by a couple of hours or so – as we get closer to that date, and the orbit continues to get better defined.) The perihelion distance is 0.914 AU (85 million miles, or 137 million kilometers), and the comet's closest approach to the earth will occur on March 22 of that year, at the relatively large miss distance of 1.315 AU (122 million miles, 197 million kilometers). This orbit suggests that Comet Hale-Bopp made its last pass around the sun about 4200 years ago, but on its way in it passes 0.77 AU from Jupiter in early April 1996; this will tend to modify the comet's orbit slightly, such that it will leave the inner solar system with an orbital period closer to 3400 years.

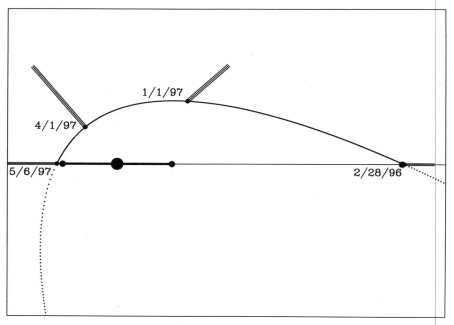

HALE-BOPP'S ORBIT AS VIEWED FROM THE PLANE OF THE EARTH'S ORBIT. The earth's orbit (seen edge-on) is indicated by the solid heavy line on the left side of the diagram.

EARLY SCIENCE AND OBSERVATIONS – THE FIRST FOUR MONTHS

For a comet to be so easily visible in an ordinary backyard telescope at a distance of 7 AU from the sun is unheard of in the history of astronomy. The only other comet which comes close to this is an object called Comet Schwassmann-Wachmann 1 (usually abbreviated as S-W 1) that travels around the sun in a near-circular orbit between the orbits of Jupiter and Saturn. Most of the time Comet S-W 1 is a dim, small object requiring a pretty large observatory telescope to be seen, but on occasion – once or twice a year, on the average – for reasons not entirely understood it experiences dramatic outbursts in brightness. During these outbursts the comet can become bright enough to be dimly visible in backyard telescopes for a week or two, before fading back into obscurity.

Some of the earliest speculation about Comet Hale-Bopp was that it was undergoing an outburst in brightness similar to those exhibited by Comet S-W 1. In fact, some of the first images obtained of the comet suggested an appearance quite similar to those of Comet S-W 1 during an outburst. If this were indeed the case, it might be that Comet Hale-Bopp really isn't an intrinsically bright comet at all; it might just be an otherwise dim comet that happened to have an outburst

COMET SCHWASSMAN-WACHMANN 1 DURING AN OUTBURST. In early 1996 this object experienced one of the strongest outbursts it has had during recent years. Photograph taken on February 19, 1996 by Takuo Kojima, and used with permission.

around the time that Tom and I found it. It might then fade back down to its "normal" brightness, and wouldn't be much of a sight to make a fuss over when it passes perihelion in 1997.

The first indication that this scenario might *not* be the case – in other words, the first indication that Comet Hale-Bopp may really be an intrinsically bright comet – came when Rob McNaught announced the 1993 image he had found. As I mentioned above, this showed the comet as a relatively bright object even when it was over 13 AU from the sun, and even at that distance it had a fairly substantial coma. For comparison, when Halley's Comet was first recovered from Palomar Observatory in 1982 it was just over 11 AU from the sun, and was an extremely dim object with no coma at all – apparently, little more than a bare nucleus. The 1993 image indicates that Comet Hale-Bopp was already a substantial object even when well past the orbit of Saturn, and rather strongly suggests that its brightness at discovery was *not* the result of any recent outburst.

In addition to this image, several other pre-discovery photographs from 1995, including at least one from as far back as late May, all show the comet near the brightness it had at discovery. And alongside all this, there has been the

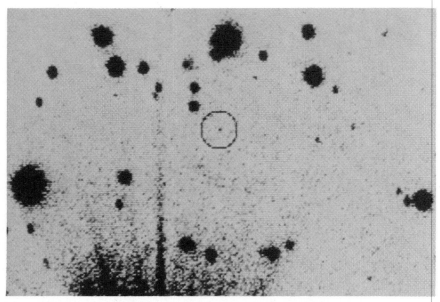

THE FIRST PHOTOGRAPH OF HALLEY'S COMET DURING ITS 1986
RETURN. *This image was taken on October 16, 1982 by David Jewitt and Gary
Danielson with the 5-meter (200 inch) telescope on Mount Palomar. Compare
this with the image of Comet Hale-Bopp (taken with a much smaller telescope)
shown on page 62. Image courtesy of the California Insititute of Technology.*

behavior of the comet itself since its discovery; not only has it not faded, it has
actually brightened, more or less as a "normal" comet. I followed it every two or
three nights from the time I found it, and by mid-October it had noticeably
brightened and was looking "better than ever."

While the possibility that a major outburst is responsible for Comet Hale-
Bopp's present brightness seems less and less likely, there is no question that the
comet's nucleus has been undergoing tremendous activity during the subsequent
months. The fact that there have been several "mini-outbursts" in the nuclear
region has been documented by careful observations by several astronomers
around the world. I've seen at least two or three such events since my discovery;
a particularly dramatic one occurred during the latter part of August, when on
the evening of the 20th I saw a bright "star" almost exactly in the center of the
coma. Since the comet has been traveling through a very "dense" star field ever
since its discovery – Sagittarius is in the very densest part of the Milky Way – I
initially thought that this was nothing more than a background star that the
comet just happened to be in front of. But, when I saw the comet again two
nights later, I saw my error; the "star" was actually a new surge of activity in the
central condensation. Over the next several nights I watched this "condensation"

spread out and fade, as the material that had been ejected from the nucleus dispersed throughout the rest of the coma.

COMET HALE-BOPP, imaged with the 3.5 meter telescope at Apache Point Observatory near Cloudcroft, New Mexico, This photograph was taken by Karen Gloria and Eddie Bergeron on October 17, 1995. Courtesy Kurt Anderson.

One question that has intrigued astronomers ever since the comet's discovery is: just what is producing all this activity? As I mentioned in Chapter 1, the primary constituent of a comet's nucleus is water ice, and usually it is the sublimation of this ice – and the expulsion of dust grains along with it – that produces most of the coma we see in a comet. But out at 7 AU, the sunlight is so weak that the temperature in space just isn't warm enough to cause the water ice to sublimate; this usually doesn't happen until a comet gets to within about 3½ AU of the sun. The most likely suspect at this distance would be carbon monoxide (CO) ice, but some of the early observations, particularly with some of the large telescopes at the European Southern Observatory in La Silla, Chile, didn't reveal any sign of this particular molecule in the comet's coma. Finally, in mid-September came the report that some astronomers, using an instrument called a submillimeter telescope located on top of Mauna Kea in Hawaii – the tallest mountain on the island, and home to some of the world's largest and most unique telescopes – had detected definite traces of the CO molecule. (Instead of being sensitive to visible light like our eyes, and most of the telescopes in the world, are, a submillimeter telescope is sensitive to extremely high-frequency radio waves.) This report has helped establish that the sublimation of carbon monoxide ice is, indeed, the main reason behind Comet Hale-Bopp's present activity and brightness.

What we see when we look at a comet's coma, especially a distant one like Hale-Bopp, is sunlight reflecting off the dust particles that have been ejected off the nucleus along with the gas. Since Hale-Bopp is as bright as it is, this would suggest that there is a lot of dust, as well as a lot of gas, in its nucleus. This seems to have been verified in a report that was released in mid-October: according to observations made with the *International Ultraviolet Explorer* – an Earth-orbiting ultraviolet telescope launched in 1978, and which has outlived its "planned" lifetime several times over – Comet Hale-Bopp produces dust at one of the highest rates ever observed in a comet. This rather strongly suggests that, if this behavior is indicative of what the comet will do when it is near perihelion, then we indeed are in for a spectacular display in 1997.

In late August some astronomers using large telescopes reported finding a bright "jet" extending from the comet's nuclear area, and spiraling outward through the coma. (Perhaps the best way to think of this "jet" is like the gush of water from a rotary water sprinkler.) Since then numerous other telescopes have seen features similar to this; perhaps the most spectacular is that seen in an image taken with the *Hubble Space Telescope* on September 26. These types of features have been seen in other comets and, just like in a water sprinkler, the spiral pattern is usually due to rotation. Astronomers thus hoped that by studying these jets over a period of time they could learn something about the nucleus' rotation; for example, how fast it rotated, what direction its polar axis points to, and so on.

Unfortunately, different people analyzing the data and using different assumptions started coming up with completely different answers, although all the analyses suggested a period of a few days. The end result is that we just don't know the comet's rotational period yet, or the direction of its north pole; it seems that the comet is reluctant to give up all its secrets just yet. In an analysis published in late October, comet scientist Zdenek Sekanina at the Jet Propulsion Laboratory concluded that more than just simple rotation is involved; it appears that the nucleus is oscillating – perhaps even "tumbling" – in a rather complex manner. Such a scenario is not unprecedented: the nucleus of Halley's Comet appeared to rotate in three different directions at three different speeds.

Tied in with all this – and perhaps one of the questions that I get asked the most, is – how big is the nucleus? Unfortunately, this is a difficult question to answer; the only actual cometary nucleus that has ever been seen directly was that of Halley's Comet, when the space probe *Giotto* flew past it in 1986 and revealed it to be an oblong object about 9 miles (15 kilometers) long in its longest direction. Since we can't see Hale-Bopp's nucleus, about all we can do is compare it with Halley under similar conditions and scale things accordingly. A lot of assumptions come into play in this process, though, so any value resulting from this is in many ways not much more than guesswork. Some of the early results suggested that the nucleus may be as much as 100 miles in diameter – about 10 times that of Halley, undoubtedly making Hale-Bopp one of the largest comets of which there's any record. On the other hand, by studying the rates at which the nucleus is ejecting gas and dust, Sekanina has concluded that it may be quite a bit smaller than this, quite possibly not much bigger than Halley's nucleus.

Perhaps the most "definitive" answers so far have come from *Hubble Space Telescope* images of the comet taken during late October. According to Hal Weaver at the Space Telescope Science Institute, analysis of these images suggest that Hale-Bopp's nucleus is no larger than 70 kilometers (about 45 miles) in diameter, with a most likely value in the neighborhood of 40 kilometers (25 miles) – a value that is still almost three times the diameter of Halley's nucleus, suggesting that Hale-Bopp may be from 25 to 30 times more massive than Halley. Weaver points out, though, that a lot of assumptions were involved in producing this figure of 40 kilometers, so we shouldn't put too much faith in it just yet.

One item that's a little less problematical is the overall size of the coma. Although the coma doesn't appear all that "big" when looked at through a small telescope, the comet's large distance from us means that even a rather small "apparent" coma can translate into a fairly large object in physical terms. The coma that we're already seeing is quite phenomenal; based on some of my own observations, the coma that's visible in a backyard telescope is about 700,000 miles across, over three-fourths the size of the sun. But, when special photographic techniques and image enhancement routines are used, the result is even more impressive: when this was done to a photograph of Hale-Bopp taken on August 18 with one of the telescopes at La Silla, astronomers determined that the coma was 1 by 1½ million miles across. Although some of this great size may have been due to the beginnings of a tail, the coma was then already quite a bit a larger than the sun, and was even larger than the coma of the Great Comet of 1811, making Comet Hale-Bopp – even at that great distance – the largest comet that has ever been seen.

Because the earth's orbital motion was beginning to carry it to the opposite side of the sun, by late October the comet was starting to be visible only low in the southwest after dusk. After the full moon in early November, Comet Hale-Bopp could only be observed with difficulty just a few degrees above the southwestern horizon, noticeably sinking lower with each passing evening. My last view of it came on the evening of the 23rd — appropriately enough, Thanksgiving evening — when I could see it only with difficulty glimmering weakly through the murky air just above the horizon. Within a couple of days the moon's presence in the evening sky obliterated the comet's weak glow and, as far as I know, no other visual sightings were made after mine. On the other hand, some observatory telescopes were able to follow it a bit longer; the latest photograph seems to have been taken at Siding Spring on December 8. And even this wasn't the last observation of the comet before its conjunction with the sun: a team of astronomers at Kitt Peak National Observatory in Arizona reported observations with a submillimeter telescope — which isn't as affected by sunlight as optical telescopes are — as late as December 10. Not unexpectedly, the comet continued to exhibit high and variable activity even at that late date.

EARLY OBSERVATIONS IN 1996

From our vantage point on the earth, Comet Hale-Bopp passed on the opposite side of the sun on January 3, 1996, and for about a month on either side of that date was invisible with any telescope we could use. During those two months, we waited; I, Tom Bopp, the world's cometary astronomers, and everyone interested in this comet wondered what would greet us when it emerged into the morning sky during February. Although it seemed unlikely, given the comet's continued brightness throughout the fall months, there still remained the possibility that what we had been seeing was the result of an extended outburst, and that we were being set up for another "fizzle" that would play itself out as the comet approached perihelion. On the other hand, if Hale-Bopp were brightening as expected, then it would be noticeably brighter when it appeared in February than it was when it disappeared in November. For many of us, then, the comet's brightness in February would be the real "acid test" as to the display we could truly expect. As January rolled on, our anxiety grew.

Because of the way the comet was situated with respect to the sun and the earth, it turned out that sky-watchers in the southern hemisphere would get a slight jump on us northerners in seeing the comet first. (I really would have liked to have been the first one to see it, but I knew that my friends in the southern hemisphere would beat me to it.) Thus, the first photograph of Comet Hale-Bopp in the morning sky was taken on February 1 by an Australian amateur astronomer, Gordon Garradd, from Loomberah, New South Wales

THE FIRST PHOTOGRAPH OF HALE-BOPP IN 1996, taken on February 1 by Gordon Garradd. Used with permission.

(near Siding Spring Observatory). Garradd's photograph indicated that the comet indeed had remained bright during the intervening months, and showed that the comet retained its rather extensive coma.

The first "visual" sightings (i.e., through a telescope) were obtained on the following morning by Arturo Gomez at the Cerro Tololo Interamerican Observatory in Chile and by Terry Lovejoy in Queensland, Australia. According to Lovejoy, who is one of the world's most experienced visual comet observers, Hale-Bopp was noticeably brighter than it had been in late November, and even exhibited a short, faint tail.

Naturally, I was ecstatic when I heard these reports, for they indicated that Hale-Bopp was indeed brightening on schedule; if anything, it was even brighter than it was "supposed" to be. (The report from Lovejoy was the first one I heard, and I spent the next few hours calling up friends and family stating "it's baa-ack!") But as exciting as this and the subsequent reports out of the southern hemisphere were, I wasn't quite satisfied; I wanted to see the comet with my

71

own eyes before I could truly believe that it was back "for real." I had to be patient, though; not only did I have to wait for the comet to become visible from the northern hemisphere, but the moon, which was full on the 4th, would be a bright object in the morning sky for the next week and a half or so thereafter and would make any attempts to observe the comet impractical during that period.

While I wanted to be the first person in the northern hemisphere to pick up Hale-Bopp, some cloudy weather here in New Mexico at the wrong time prevented it. As far as I know, the first person to see it was an amateur astronomer on the Mediterranean island of Malta, Umberto Stagno, who successfully picked it up on the morning of February 13. But he didn't beat me by much: on the following morning – Valentine's Day, of all days – I did manage to see it, low in the southeastern sky around the beginning of dawn. Despite its low altitude above the horizon and the bright background sky – both of which tend to decrease the contrast of the comet with its background and thus make it harder to see than it otherwise would be – I could easily tell that it had brightened significantly since November, and I was able to verify to my own satisfaction that the reports by Lovejoy and others were correct. Hale-Bopp, indeed, was back!

HALE-BOPP ON APRIL 18, 1996. This image shows the beginning of a tail that is starting to become apparent. Photograph taken by Tim Puckett, used with permission.

As the comet continued to rise higher and earlier during the subsequent days many other astronomers throughout the world were able to pick it up with their telescopes. Tom Bopp got his first view of it on Sunday morning, the 18th, from the same site in the Arizona desert where he had discovered it the previous July and, like me and everybody else, was excited to see how our comet had brightened during the intervening months.

One of the first scientific studies of the comet after its reappearance from behind the sun was a set of observations with the submillimeter telescope on Mauna Kea on February 10. These indicated that Hale-Bopp is continuing to produce large amounts of carbon monoxide, at a rate at least equivalent to that detected the previous fall. The astronomers involved estimate that during the time since its discovery the comet has ejected up to 10 million tons of carbon monoxide; while this may seem like a large number, this is in actuality a very tiny fraction (probably one-billionth or less) of the comet's total mass.

As I write these words (mid June), the comet has continued to brighten even more than expected. It has reached the point where it can be seen with ordinary binoculars and small telescopes if one knows where to look, and several experienced amateur astronomers have reported sightings of the comet with their naked eyes. A recent report from Jacques Crovisier at the University of Paris suggests that water from ice sublimation in the comet has now begun in earnest. The signs thus continue to look promising!

While almost anything can still happen with Comet Hale-Bopp and the display it will put on, all these observations suggest that the great show we've all been hoping for has a reasonable likelihood of happening. At the very least, we have some assurance now that the comet will become a moderately bright naked-eye object in our skies next spring. Readers should always keep in mind the "fickle" nature of comets, but I believe we have real cause for optimism now, and in my discussions in Chapter 3 I will present what I consider to be a reasonably realistic scenario.

METEORS FROM HALE-BOPP?

Recalling my discussion in Chapter 1 about the relationship between comets and meteor showers, one might ask if there is any possibility of seeing a meteor shower from Comet Hale-Bopp. The comet's orbit does pass somewhat close to the earth's orbit; the closest separation between the two orbits is a little over 0.1 AU (approximately 10 million miles), and the comet will pass through this point on May 6, 1997, the earth being nowhere near at that time. The earth is at this "point of closest approach" around January 3 of each year, and any meteor shower we might see resulting from Comet Hale-Bopp would be seen around that date.

As it so happens, the Quadrantid meteor shower peaks every year on or about January 3. (This shower is named for the now-no-longer "recognized" constellation Quadrans Muralis – the Mural Quadrant, an old instrument once used for measuring stars' positions – which is now a part of constellation Boötes, southeast of the Big Dipper's handle.) The Quadrantids are one of the strongest meteor showers that the earth encounters during its annual trip around the sun, and have during some recent years produced meteors at the rate of 100 to 150 per hour. Furthermore, no parent comet for the Quadrantids has been positively identified yet. Is it possible, then, that Comet Hale-Bopp is the parent comet for the Quadrantid meteor shower?

As one might suspect, that question has been batted around by astronomers ever since the true nature of Hale-Bopp's orbit became apparent. Despite the coincidence of the dates, it looks like the answer to the question is a reasonably firm "no." The orbit of the Quadrantid meteor shower is pretty well known, and while there are some similarities between this and Comet Hale-Bopp's orbit, there are enough differences to suggest rather strongly that there is no relationship between them. Among other things, the orbital inclinations are quite a bit different (about 72° for the Quadrantids, versus almost 90° for Hale-Bopp), and the orbital period for the Quadrantids is in the neighborhood of 5 to 7 years, as opposed to the several thousand years for the comet. Some astronomers have pointed out that some potential parent comets for the Quadrantids have indeed been identified, although at this time none of these identifications are certain. In addition, the expected point in the sky from which any Hale-Bopp meteors might originate has been calculated to be in the constellation Corona Borealis which, although somewhat close to the originating point for the Quadrantids, is still quite a bit to its south.

While the "miss distance" of 0.1 AU between Hale-Bopp's orbit and the earth's orbit is considered by most astronomers to be too large to produce any significant meteor shower, the potential for observing Hale-Bopp meteors nevertheless does exist, and would make a worthwhile observing project. Keeping in mind the Leonids (see Chapter 1), which peak a year or so after the return of their parent comet, it would seem that the best chance of observing a Hale-Bopp meteor shower would occur in 1998 (or perhaps 1999). On the other hand, it might also be worthwhile to check for this during the preceding years as well. To my knowledge, nothing unusual was noted at the appropriate time in 1996, although the nearly full moon at the time may have contributed to this result.

AND THE MISINFORMATION FLIES . . .

As I pointed out in Chapter 1, for the past many centuries humanity has tended to associate the appearance of a bright comet with various calamaties

Top astronomer's grim warning...

COMET HEADS FOR GIANT CRASH WITH EARTH

A GREAT COMET discovered just a few months ago is hurtling through the solar system on a collision course with Earth – threatening the very existence of humanity, a top astronomer warns.

What's more, the "Hale-Bopp" comet could crash head-on into our planet on April Fools' Day next year – ending the world with a cruel joke.

"The comet was strikingly visible while passing through the orbits of Jupiter and Saturn," says Johan Oftebrau, a noted astronomer from Oslo, Norway.

Enormous

"Since it's so bright from that distance, it shows how enormous it is. And it's definitely going to come close enough to crash into earth.

"This is not the first time a comet has come so close. In 1910, the Earth actually passed through the tail of Halley's Comet.

"It is also believed that a comet crashed into the Earth millions of years ago causing the extinction of dinosaurs from its fallout.

"I'm afraid...very afraid ...the path of this comet will have similar consequences. It could be devastating."

Experts have predicted the comet – discovered last July by U.S. observers Alan Hale and Thomas Bopp – will be at its brightest on April 1, 1997.

"Unfortunately, that's

●*JOHAN OFTEBRAU: Sees disaster ahead*

based on the conclusion that it will pass more closely to the sun," says Dr. Oftebrau.

"I believe now its path will bring it closer to Earth based on my intensive and extensive calculations – and strike our planet on that date."

Freakish

If the comet does continue on its apparent collision course with Earth, Dr. Oftebrau says, the first signs will occur a few months before, perhaps in January 1997.

"It will be far more devastating than the freakish weather we've been experiencing in recent years," says Dr. Oftebrau.

"There will be hail-

●*THE COMET will have a devastating effect – much like the blast of hundreds of nuclear missiles*

'Doomsday' prediction: It just cannot miss!

stones the size of canned hams from the comet's ice, covering every bit of land and water on the Earth.

"Following that will be enormous flooding and major earthquakes be- cause of the Earth's shift-

ing caused by the coming comet. It will get gloomier and colder as the comet gets nearer.

"There will soon be total darkness and freezing temperatures everywhere – then the explosion of impact," Dr. Oftebrau be-

lieves there's a possibility of some survivors, but the world will be left in darkness and coldness because of fallout from the comet blocking the sun's rays.

"This will last for

decades. The survivors, clearly smarter now than the prehistoric dinosaurs, will have to use all their modern ingenuity to rebuild civilization," he predicts.

"Of course, I pray my calculations are wrong ...but I've been a scientist for a long time."
– *FRED SLEEVES*

"A popular song is one that makes us all think we can sing." – ANDY WILLIAMS

16 — SUN — January 9, 1996

AN EXAMPLE OF TABLOID JOURNALISM.

occurring on the surface of the earth, e.g., earthquakes, wars, epidemics, etc. Usually this association has been so strong that the particular comet in question has been thought to foretell – in some cases, even "blamed" – for the particular calamity. This has happened notwithstanding the fact that such disasters occurred all the time, regardless of whether or not a bright comet happened to be in the sky at the time of the event in question. It just seems to be a part of the human psyche to associate events in space with events on the ground and, if possible, to try to foretell future events on Earth by examining the goings-on in space. This is what led to the practice of astrology, apparently begun in ancient Babylonia several thousand years ago.

Old habits die hard, and despite our supposedly more enlightened society today, there still remains the commonplace desire to associate terrestrial happenings with events in the sky. Just witness the prevalence of daily astrology columns – practically every newspaper in the country has one – and the frequency with which questions like "What's your sign?" are asked. Despite numerous studies showing no correlation between planetary positions in the sky and events on the earth, the will to believe remains quite strong among some segments of the public.

Along these lines, the appearance of a bright comet like Hale-Bopp (or, at least, the Hale-Bopp display we hope to see) is sure to produce numerous predictions of future – usually dire – events. (There were certainly no shortages of these prior to the appearances of Comet Kohoutek in 1973 and Halley's Comet in 1986.) In the case of Comet Hale-Bopp, we may see a dramatic increase in this type of activity, coming so close as it does to the end of the 2nd Millenium. Many "end-of-the-world" predictions are already starting to be made – and such will certainly increase – because of this timing. I've already seen or heard suggestions that the appearance of Hale-Bopp is contained within the prophecies of Nostradamus – and presages some particularly troubling episodes in the earth's history – and may also be contained within the prophecies of some of the Native American tribes. I've also seen statements indicating that Hale-Bopp may be one of the "signs of the end times" and may even be the star "Wormwood" mentioned in Chapter 8 of the Biblical book of *Revelation*.

As I've tried to indicate throughout this book, there is no reason to believe – and no evidence that suggests – that the presence of a bright comet in the inner solar system has any effect whatsoever on any events occurring on the earth (other than making us astronomers lose sleep while we're studying it). The fact that Hale-Bopp is appearing so close to the end of the millenium may be interesting to some, but really is not so dramatic as it might seem; since naked-eye comets do appear about once a year, on the average, and Hale-Bopp's appearance is almost four years before the end of the millenium, there's a very good possibility that at least one other naked-eye comet will grace the skies after Hale-Bopp's appearance but before the millenium is out.

(If one thinks about it, the end of the 2nd Millenium is itself a rather arbitrary point in time. Although our current calendar is based upon the time of the birth of Christ – and perhaps we should keep in mind that a major portion of the earth's population does not subscribe to a belief in Christianity – this chronology was actually worked out by Dionysius Exiguus in the 6th Century A.D. Since there are no firm historical records as to the exact time of Christ's birth – some of the New Testament clues being inconsistent by up to a decade or more – Dionysius could do little more than make an educated guess as to when this occurred.)

Keeping all this in mind, then, I'd like to go on record as stating that Comet Hale-Bopp has absolutely no relationship to any events on the Earth's surface, and does *not* foretell any dire or catastrophic events that might take place in our near future. There may well occur such events as wars, earthquakes, etc. during and after the comet's appearance, but these would happen regardless of whether or not the comet had appeared; I ask the reader to name any recent years when such events did *not* occur. Comet Hale-Bopp should, we hope, be a temporary and spectacular addition to our nighttime sky, but that is *all* it will be.

Besides this business about prophecies and foretelling of future events, there are various other forms of misinformation about Hale-Bopp making the rounds now, and these, of course, are sure to increase despite my efforts in this book and those of other astronomers elsewhere. A classic example was a segment about the comet that recently aired on the FOX-TV show "Sightings" which begins with the incredible statement that comets are "enormous balls of nuclear energy hurtling through space destroying everything in their path" and then goes downhill from there. About the only correct information given in the program was the parting statement that Hale-Bopp is "not expected back for another 3000 years." Many of the various statements that are circulating are really not worth responding to here; for example, the claims that Hale-Bopp is an alien mothership, or is being controlled by aliens as a distraction to some "real" threat, or is a non-existent object concocted as part of some government conspiracy – all of which I've heard during the months after the comet's discovery. It's an unfortunate state of affairs that, over fairly large segments of our society, people are going to believe whatever they want to believe, and anyone who subscribes to and promotes ideas such as these is not going to be swayed by anything I say here.

There are, nevertheless, a couple of items being thrown around that I'd like to address. I've seen some discussion to the effect that the comet has exhibited "course corrections" since its discovery; such statements have usually been made in the context that the government has known about the comet for some time, that the comet has made these "course corrections," and that the government is trying to hide this fact. Such behavior is impossible for a natural object in the solar system to exhibit; all such objects obey Newton's law of gravity, and their motion can be predicted within this law. To reiterate something I discussed earlier, the *calculated* orbit of an object like a comet may change over time as new data comes in and the calculations are refined, but the *true* orbit itself does not change. Such was the case with Comet Hale-Bopp after its discovery; recall from my discussion above that it took awhile for the nature of the comet's orbit to be revealed, and that some of the earliest calculated orbits didn't necessarily pan out into "reality," but this was a function of the data we had available, not of the orbit itself.

Is so happens that a comet's orbit does, in fact, change by (usually) small amounts, primarily as a result of the gravity of a planet such as Jupiter. Also, the eruptions of gas and dust off a comet's nucleus can act as small rocket engines, and these can cause a comet's motion to deviate slightly from its predicted orbital path; this effect is usually described under the term "non-gravitational forces" by cometary astronomers. All of these effects, however, are understood and can be incorporated into the calculations of a comet's orbit; none of them constitute the effects one might associate with the term "course corrections," which implies something unforeseen and unpredictable.

Finally, let me address the issue of the comet's colliding with the earth. We have every reason to believe that the orbit we have available now is a reliable one and, according to this orbit, there is absolutely *no* chance that Comet Hale-Bopp will collide with the earth. Repeating my earlier discussion, the comet's closest approach to the earth will occur on March 22, 1997, and at that time it will still be 1.31 AU away; in other words, the earth is closer to the sun than it will ever be to the comet. It is true that the comet's orbit comes to within about 0.11 AU of the earth's orbit, but that point is near the position the earth occupies in early January, whereas the comet will be at that point in early May.

Before the comet's orbit could be determined with a believable degree of reliability, no astronomer could state with absolute certainty that there would be no collision; the possibility could not be completely excluded as long as the orbit remained uncertain. Some of the more sensationalist-oriented elements of the press latched onto this – quite irresponsibly, in my opinion – and implied that this meant a collison was a distinct possibility, and that astronomers were "wavering" on this prospect. The truth was in fact nothing of the sort; there is a big difference between an event's being *theoretically* possible – for example, flipping a coin 100 times, and having it turn up "heads" each time – and its being a possibility that we have to concern ourselves with in any real way.

THE PAST AND FUTURE OF COMET HALE-BOPP

We know from the present orbital calculations that Comet Hale-Bopp has been around at least once, a few thousand years ago, but we don't know how many times before that. If we could determine exactly when that last return occurred, which would then allow us to determine where the planets were at that time and calculate their respective effects on the comet's orbit, it might be possible to estimate the comet's history before that, but unfortunately this can't be done. Similarly, we can't tell exactly when Hale-Bopp will make its next return, so we can't make any specific predictions as to its returns after that. One thing we *can* do, though, is calculate the effects of the planets' gravity on the comet's orbit in a statistical sense, and see if there are any long-term trends in that orbit.

Such a set of calculations has recently been performed by Mark Bailey at the Armagh Observatory in Northern Ireland and several of his colleagues in Europe and America. By examining the comet's orbit over a period of about five million years centered on the present, they find that the orbital period may have been quite a bit shorter in the past, perhaps as short as a few hundred years. It is even more likely that Hale-Bopp could have an orbital period this short, or even shorter — perhaps not drastically different from that of Halley's Comet — sometime within the next few hundred thousand years. Furthermore, there is a chance that at some point in the future, Comet Hale-Bopp's perihelion distance will be quite a bit smaller than it is now, and consequently our descendants in some far-off future era may see a display that's even more spectacular — perhaps very much so — than the one we'll see in 1997.

Since the comet's orbit does come relatively close to the earth's orbit, this would suggest a possibility that, at some point in time in the distant future, a collision with Earth may occur. (It also suggests that some past and future returns may have been or may be extremely spectacular.) On the other hand, the comet's orbit on its inbound track passes even closer to Jupiter's orbit which, besides making Jupiter the dominant player in affecting its orbit, suggests that a collision with that planet is a distinct possibility; it certainly is more likely than any collision with Earth. Perhaps our descendants at some point in the distant future may witness a higher-scale version of the Shoemaker-Levy 9 impacts.

It is tempting to think that records of Comet Hale-Bopp's most recent return might exist within the writings of some ancient civilization. Indeed, when the early orbits were suggesting that this return occurred about 3000 years ago — in other words, somewhere around 1000 B.C. — there was some discussion being made that the comet's appearance might be identified in old Chinese or Babylonian records. Unfortunately, the more recent calculations indicate that the previous return occurred somewhat further in the past than that, probably somewhere around 2200 B.C. At that time, there doesn't seem to have been any civilizations where the maintaining of such records was taking place; the only ones that might have done so were the Egyptians and the Sumerians. But even if such records were maintained, almost certainly none of them would exist today to be examined, and even if there were, any descriptions of the comet would probably be so hopelessly intertwined with the mythologies of the time that it would be all but impossible to extract any meaningful information from them.

On the other hand, the present orbital calculation indicates that the comet's next return should be around A.D. 5400, and it is certainly an interesting question to ask how precisely we might be able to predict when that will be. Unfortunately, it is not a straightforward calculation; a lot depends on the strength of the non-gravitational forces that affect the comet's trajectory.

Usually, it takes two returns of a comet to include these properly into an orbital solution, and of course we don't have that second return to work with. Brian Marsden tells me, though, that if the world's observatories can follow Comet Hale-Bopp long enough after perihelion, it should still be possible to predict its next return to within about a decade. There is, in fact, a good possibility this might occur; Halley's Comet was followed for almost eight years after perihelion, the last observations being obtained in January 1994, when it was almost 19 AU from the sun. Based upon its current brightness and that exhibited in 1993, it would seem that Hale-Bopp should be followed until the middle of the next decade or longer, at which time it would be some 20 AU or more from the sun. It is a most refreshing thought to me to think that our descendants at that far-off time in the future might see this comet, and know that it is the same Comet Hale-Bopp that appeared at the end of the 20th Century.

CHAPTER 3: PROSPECTS FOR COMET HALE-BOPP

To my mind, the only civilized technique for observing comets requires that rare object – one bright enough to allow lounge chair viewing from the back yard.
— Don Yeomans*

COMPARISONS: HALLEY'S COMET

By the end of the summer in 1995 we knew the orbit of Comet Hale-Bopp relatively well; we know where it's going to be in 1997, and where and when to look for it. The one thing we don't know, though, is just how bright the comet will get and how spectacular it will look. As I've tried to stress repeatedly throughout this book, about the only thing that's predictable about a comet's brightness is its unpredictability; to reiterate my expression from Chapter 1, as far as Hale-Bopp is concerned, "the comet's going to do whatever the comet's going to do." While the comet's appearance in early 1996 does give us reason to be optimistic, any statements as to how the comet will continue to perform still can't be much more than educated speculation. Keeping this in mind, then, I'll go ahead and take a stab at predicting just what "the comet's going to do."

One thing which makes this business of predicting Comet Hale-Bopp's brightness so difficult is the lack of appropriate precedents. Most of the bright comets which occasionally grace our skies aren't discovered until they are fairly close to the sun, and thus we don't get much of an idea of what they're like when they are as far from the sun as Hale-Bopp was when it was discovered. At the same time, while comets occasionally have been discovered at that distance, almost without exception these have been relatively dim objects which have drawn no closer to the sun. Only once before was a comet found at such a distance and subsequently came in fairly close to the sun and Earth. This was the most recent return of Halley's Comet in 1986, where we had the advantages of knowing it was on its way in and of knowing where to look, and accordingly we were able to pick it up well ahead of time. Along these lines, then, let's take a quick look at that return and see what comparisons we can make with Comet Hale-Bopp.

At its most recent return Halley's Comet was first photographed in October 1982 with the 5-meter (200 inch) Hale reflector at Palomar Observatory in California. (This telescope was named for the early 20th Century astronomer George Ellery Hale, not for me!) At that time it was slightly over 11 AU from the sun, and appeared as nothing more than an extremely faint star-like object with no trace of a coma. Even two years later, when its distance from the sun

---

* Don Yeomans, of the Jet Propulsion Laboratory in Pasadena, California, is considered one of the foremost experts on cometary orbits in the world.

had shrunk to 6 AU, it could only be photographed through large telescopes, although by this time it was at least starting to exhibit a tiny coma. It was first sighted visually through a telescope in January 1985, by which time it had closed almost to within 5 AU of the sun; but this observation was made by an extremely capable and keen-eyed observer looking through a large telescope located in the crystal-clear thin air at the top of Mauna Kea (elevation almost 14,000 feet) in Hawaii, and even then the comet appeared as nothing more than a tiny dim smudge, barely distinguishable from a star.

Halley's Comet finally became bright enough to be visible to experienced amateur astronomers with decent telescopes by about the latter part of August 1985, it then being about 2.8 AU from the sun. After that it brightened rapidly, becoming visible in a pair of binoculars by about the end of October, and to the naked eye (of an experienced observer located in a dark environment) by about the middle of November; at that time it was approximately 1.7 AU from the sun (but only about 0.7 AU from the earth, and located on the opposite side of the earth from the sun, both of which helped to make the comet brighter and easier to see.) After November Halley began to pull away from the earth, although it continued to brighten slowly as its distance from the sun decreased. By mid-January 1986 it had closed to within 0.8 AU of the sun, and people located in a dark environment and who knew where to look could see it, with a short tail, in the evening sky with their naked eyes. At that time the comet was passing over to the opposite side of the sun as seen from the earth, though, and thus it was no longer visible after the end of January.

At perihelion on February 9, Halley's Comet was 0.59 AU from the sun and 1.55 AU from the earth, and invisible as far as we Earthlings were concerned. Toward the end of February it became visible in the morning sky, about as bright as some of the brighter stars in the Little Dipper. It maintained this brightness throughout most of March, as its distance from the sun increased while at the same time it was approaching the earth. To me, Halley's Comet was a pretty nice sight to the naked eye during this time, with a moderately nice and bright tail, although it was a far cry from the "Great Comet" that Comet West had been ten years earlier. The comet was actually brightest during the early part of April, when it was located some 1.3 AU from the sun but only 0.4 AU from the earth; at that time it was about the same brightness as the stars in the Big Dipper, but because of its location beyond the earth the tail was directed almost directly away from us, and the comet appeared as little more than a fairly large and bright diffuse cloud. After that Halley's Comet faded quite rapidly, although experienced amateur astronomers were able to follow it for several more months, and professional astronomers at large observatories were able to photograph it well into the 1990s.

If Comet Hale-Bopp were to exhibit a change in brightness proportional to that of Halley, then the prospects for a spectacular display in 1997 are very good indeed. Throughout the parts of its orbit where it has been observed so far – the photograph from Siding Spring in 1993, its discovery in 1995 and the months thereafter, and its appearance in the morning sky in early 1996 – Comet Hale-Bopp has consistently been several hundred to a couple of thousand times brighter than Halley was at a similar distance from the sun. *If* it were to maintain this trend up through the time of its perihelion passage, Hale-Bopp might well become visible to the naked-eye by the middle of 1996, be a fairly bright object by the end of the year, and by the time it passes perihelion in early April 1997 it might very well be quite a bit brighter than the planet Venus – the brightest object in our sky apart from the sun and the moon – and could even be visible to the naked eye during broad daylight. It might continue to be visible to the naked eye in the nighttime sky for several months thereafter, possibly up through the end of 1997.

This is certainly an optimistic scenario, and I would caution against putting much faith in it. At the same time, while many astronomers – quite rightfully, in my opinion – might say this type of scenario is "unreasonable," I would point out that for a comet 7 AU from the sun to be visible in an amateur astronomer's modest telescope is also fairly "unreasonable." We're just going to have to wait and see for ourselves what the comet does.

COMPARISONS: THE GREAT COMET OF 1811

If one looks through the records of comets which have been observed during the past few centuries, one object stands out as being remarkably similar to Comet Hale-Bopp in many of its orbital and physical characteristics. This is the Great Comet of 1811, which I've already mentioned as having exhibited the largest cometary coma ever observed – at least, until Comet Hale-Bopp came along – and which was a spectacular object in the nighttime skies of that long-ago year. It was bright enough to be included in some of the popular literature of the time – including, as I've noted before, in Tolstoy's *War and Peace* – and several segments of the population at the time thought there might be links between the comet and Earthly events, including earthquakes in South Africa and the New Madrid area in southeastern Missouri, as well as some apparently exquisite wine vintages from Portugal and France. In addition to their large coma sizes and high intrinsic brightnesses, both comets have relatively large perihelion distances (1.04 AU for the comet of 1811, vs. 0.91 AU for Hale-Bopp), stayed relatively far from the earth during their respective appearances (the closest approach distance was 1.22 AU for the comet of 1811, vs. 1.31 AU for Hale-Bopp), and have orbits steeply inclined to the Earth's orbit (the inclination of the comet of 1811's orbit was 107° – i.e., 73°, but in the opposite direction of the earth's orbit – vs. 89° for Hale-Bopp). The orbital periods of the two comets are quite similar, both being in the approximate range of 3000 years.

THE GREAT COMET OF 1811, as it appeared over the skies of England in October of that year. Courtesy Fred Whipple.

Despite the many similarities, there doesn't seem to be any reason to believe that the two objects are in any way related to each other, and thus we shouldn't draw too many conclusions about what to expect from Comet Hale-Bopp based upon the earlier comet. Nevertheless, it can be instructive to examine just what the comet of 1811 did, with the understanding that this can be nothing more than a rough guide to the performance of Hale-Bopp in 1997.

The Great Comet of 1811 was discovered on March 25 of that year by a French astronomer, Honoré Flaugergues; at that time it was already dimly visible to the naked eye, despite being over 2.7 AU from the sun and almost six months away from perihelion (which occurred on September 12). It brightened steadily over the next few months, and the astronomers of that time followed it up until mid-June, when it disappeared on the far side of the sun. It was picked up in the morning sky during the latter part of August, still visible to the naked eye although not all that bright, and it maintained this brightness up through its perihelion passage, at which time it was still 1.6 AU from the earth. Although receding from the sun after that, the comet continued to approach the earth, with closest approach occurring about October 16. As this occurred it brightened dramatically, while at the same time being located far in the northern part of the sky, and thus was easily visible to people in the northern hemisphere. In mid-October the comet was described as being as bright as the brightest stars, and as possessing two long tails, one of which was curved and was apparently 7° — fourteen times the apparent diameter of the moon — *wide*. While the comet faded in brightness after that, the tail continued to grow, until by early

December its true length was about 1.3 AU, and according to one report it stretched across 70° of sky (roughly ¾ of the horizon-to-zenith distance).

In January 1812 the comet was still visible to the naked eye, although nowhere near as bright as it had been a couple of months earlier. At this time it was also approaching the sun (as seen from the earth's vantage point), and Earthbound viewers lost sight of it by about the middle of the month. Astronomers picked it up again in mid-July, at which time it was only visible through a telescope – although it did sport a short tail even then – and followed it for another month before losing sight of it altogether. If the astronomers then had had access to the kind of telescope equipment we have today, we can be sure they would have followed it for a much longer time than that, but even so the almost 17 months of visibility – almost nine of those months with the naked eye – set a record for that time, which wasn't surpassed until several decades later when better equipment was available.

In addition to the many similarities between the two comets, the viewing geometries of their repective appearances are also quite similar. When nearest the sun and the earth both comets are well north of the earth's orbital plane, and are visible in the far northern sky (and thus easily accessible to viewers located in the northern hemisphere). The geometry for the comet of 1811 is slightly the better of the two, in that when at its best its apparent separation from the sun was larger than Hale-Bopp's will be; this means that it was higher up in a darker sky. This allows a better contrast with the background sky, and also means its light doesn't get dimmed as much when it passes through our atmosphere. (To see what I'm talking about here, try following a bright object – such as a bright star or planet – as it sinks lower in the western sky before it sets. It will appear to grow dimmer as it gets lower; its light has to travel through more atmosphere to get to us, and thus more of its light gets absorbed or scattered off.) Comet observers in 1811 also got a slightly more "broadside" view of that comet's tail than we will get of Hale-Bopp, meaning that, if Hale-Bopp's tail were to be the same length (in physical terms) as that of the comet of 1811, it wouldn't appear as long in our sky, since we would be seeing it somewhat foreshortened.

Despite the fact that the differences I've just mentioned work against Hale-Bopp a little bit, these differences are not all that significant, and won't diminish our view of the comet in any major way. If Comet Hale-Bopp does perform in a matter similiar to that of the Great Comet of 1811, I think we are all in for a good show; after all, it wasn't called the "Great Comet" of 1811 for nothing. If Hale-Bopp lives up to this potential, it may well be referred to in later years as the Great Comet of 1997 although, for obvious reasons, I'm still partial to the name Hale-Bopp!

COMPARISONS: COMET HYAKUTAKE – THE GREAT COMET OF 1996

On the morning of January 30, 1996, a new comet was discovered by Yuji Hyakutake, a Japanese amateur astronomer who resides near the southern tip of the island of Kyushu. (This was Hyakutake's second comet discovery within a period of five weeks, incidentally.) At its discovery this newer Comet Hyakutake was a rather dim object located within the constellation of Libra, the scales, but once the first orbital calculations were performed a few days later these indicated that we were in for a treat. Perihelion was to occur on May 1, at the rather small perihelion distance of 0.23 AU – roughly similar to that of Comet West in 1976 – and, moreover, the comet would pass only 0.102 AU (9½ million miles) from the earth on the night of March 24-25. Subsequent observations soon showed that Comet Hyakutake had an intrinsic brightness and activity level comparable to that of Halley's Comet, and refined orbital calculations soon showed that the comet was not a new one from the Oort Cloud, but last appeared in the inner solar system some 8000 years ago.

COMET HYAKUTAKE ON MARCH 25, 1996, the night of its closest approach to the earth. The stars of the Big Dipper's handle are at the upper right. Photograph by the author; 2-minute exposure of ASA 1000 film.

Comet Hyakutake brightened rapidly, and by the end of February experienced astronomers from dark sites were already able to spot it with their naked eyes. By mid-March, after the moon had cleared out of the morning sky, the comet had brightened to the point where many comet watchers were able to pick out several degrees of tail in addition to the overall coma. And by the night of its closest approach to the earth Comet Hyakutake was as bright as the brightest stars in the nighttime sky, and I'm sure many readers will remember it as a truly awe-inspiring sight riding high in the night. While the comet began to fade somewhat in the days after that as it pulled away from the earth, we began to get a more broadside view of the tail – which until that time had been more or less directed away from us – with the result being that we could observe it as being enormously long. For example, on the morning of March 27 I was able to see the tail stretch across 70° of sky, and some other astronomers were reporting it as being even longer than that.

As expected, the comet continued fading as it traveled further from the earth, although it remained relatively easy to see with the naked eye. After about mid-April, by which time it was starting to appear low in the northwestern sky after dusk, it was expected to commence brightening again as it approached perihelion, but for some reason this didn't happen, at least not to the extent that had been anticipated. As I write these words in late April, Comet Hyakutake has all but disappeared into the evening twilight, and it looks like the bright show we were expecting about this time won't materialize. For us northern hemisphere observers, the comet is gone; after perihelion it will only be visible from the southern hemisphere, and unless something dramatic happens, it will probably not be too bright then anyway.

In a way, this failure to brighten prior to perihelion could be considered a "fizzle," although a low-grade one, to be sure. What could have caused this? As it turns out, images taken with the *Hubble Space Telescope* (and with several Earthbound telescopes as well) when the comet was near Earth showed what were apparently tiny fragments within the coma that appeared to have recently broken off the nucleus. This fragmenting could have exposed more material in the nucleus to sunlight, in turn causing a (temporary) increase in brightness. (Recall the discussion of Comet West in Chapter 1.) Fortunately for us, this occurred when the comet was about 1 AU from the sun, when it just happened to be close to the earth. It is quite likely that this fortuitous set of circumstances contributed to the overall splendor of the show we witnessed in late March.

Despite its mediocre performance in April, Comet Hyakutake's grand appearance when it passed by the earth most definitely warrants it the nickname of the "Great Comet of 1996." There then follows the very valid question – and it's one which I'm getting asked quite a bit – as to how Comet Hale-Bopp will compare with this display just given us by Comet Hyakutake. To answer that, we need to look at how the two objects are alike and how they are different.

First off, Hale-Bopp is a *much* larger and brighter comet intrinsically than is Hyakutake; if the two comets were to be viewed under identical circumstances Hale-Bopp would far and away outshine the other one. On the other hand, Hyakutake's perihelion distance is quite a bit smaller than that of Hale-Bopp, and of course Hale-Bopp will come nowhere near as close to the earth as Hyakutake did. (Hale-Bopp's closest approach to the earth is almost thirteen times further away than was Hyakutake's.) Furthermore, the viewing geometry for Hale-Bopp will not be as good as it was for Hyakutake. Recall that at its closest approach to the earth Comet Hyakutake was visible high up in a dark sky for most of the night; Hale-Bopp, on the other hand, will not be so well placed. The geometry for viewing Hale-Bopp's tail will also be less favorable than was that for Hyakutake.

What this all comes down to, essentially, is this: assuming that Comet Hale-Bopp keeps on brightening as it should – and at this point we have good reason to believe that it will – then during the spring of 1997 it should still be quite a bit brighter than was Hyakutake during the spring of 1996. We will not have the opportunity to see Hale-Bopp high in a dark sky like we did with Hyakutake, though, and this will detract from its visibility slightly. And while Hale-Bopp may grow a tail that is quite a bit longer in physical terms than was Hyakutake's, it is quite unlikely that it will appear as long in our skies as Hyakutake's tail did. That tail might be brighter than Hyakutake's tail, however, and usually it isn't just the length of the tail, but how bright that tail is, that determines how spectacular a comet appears. One more point in Hale-Bopp's favor is that, while Hyakutake's appearance in our skies was rather brief, Hale-Bopp's sojourn will last for several months, affording all of us a multitude of opportunities for viewing it.

The appearance of two very bright comets in rapid succession like this is quite unusual, but of course is nothing more than coincidence. In many ways it's quite fortunate, since the excitement generated by Comet Hyakutake's appearance should make that of Comet Hale-Bopp even more meaningful for us comet-watching Earthlings. (I've been going around saying that Comet Hyakutake is an "appetizer" for the "real show" that's coming next year.) It is even possible – although I hate to admit it – that Comet Hyakutake could end up being the more spectacular of the two objects, and it is certainly true that Hale-Bopp will have a hard act to follow. But I have hope, and readers can rest assured that during a couple of the late March mornings while I was watching Comet Hyakutake glisten in the northern sky I was also looking at Comet Hale-Bopp through my telescope and giving it some rather enthusiastic pep talks.

THE APPEARANCE OF COMET HALE-BOPP: AN OVERVIEW

As I write this the comet has now appeared in the morning sky, and is visible in the southeast before dawn; experienced observers who know where to look can pick it up in a small telescope, or even with a pair of decent binoculars. By the time this book is available – which should be about by late summer 1996 – the comet will be near *opposition*; i.e., located in a position such that, as seen from the earth, it is directly opposite from the sun. (In other words, it rises around sunset and sets around sunrise.) If the comet has continued to brighten somewhat "normally" during the interim, by this time it should be easily visible in a small pair of binoculars. As 1996 progresses Comet Hale-Bopp moves over into the evening sky, traveling north against the background stars as it does so, and during the latter months of the year it will be visible in the west during the hours after dusk. Sometime during this period the comet should start to become visible to the naked eye, at least to people who are located in dark rural sites. By the end of 1996 the comet will be almost directly north of the sun, and will be hard to observe; if it's bright enough, however, it might be seen both low in the northwest during dusk and low in the northeast before sunrise.

In early 1997 Comet Hale-Bopp becomes visible in the morning sky, now moving to the north fairly rapidly, and (presumably) becoming much brighter. By early March it may well be a spectacular object in the pre-dawn sky, very possibly as bright as the brightest stars (if not brighter). As March progresses the comet will start to sink lower toward the northeast horizon, and just after the spring equinox it will once again be directly north of the sun. By this time, though, it will be far enough away from the sun, and hopefully bright enough, that visibility in both the evening and morning skies is possible. From far enough north – say, a latitude of 45° north or higher – the comet will not even set, but will stay above the northern horizon all night.

The best time for viewing the comet from the northern hemisphere will be in the weeks after perihelion passage in early April. At that time it will be visible in the evening sky, in the northwest after dusk. Theoretically, it should also be at its brightest, and may be exhibiting its longest tail, during this period. As April extends into May, the comet will rapidly be moving southward, and as a consequence sinking lower and lower close to the western horizon after sunset; it should also start to fade somewhat. After the latter part of May the comet will probably not be visible from the northern hemisphere; at the same time, observers in the southern hemisphere should be able to see it near the western horizon during the early evening hours.

In early July the comet will be directly south of the sun, and if it is still bright enough, our friends in Australia, South America and southern Africa may be able to see it in the southwest after sunset and in the southeast before sunrise. In the months after that it will rise higher in the morning sky as seen from

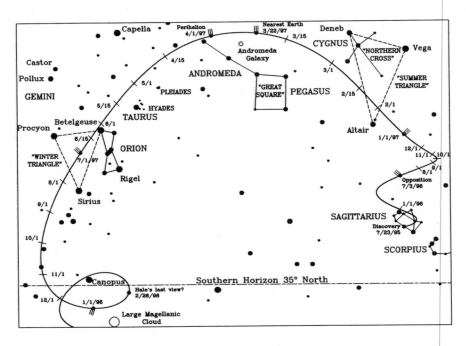

PATH OF COMET HALE-BOPP THROUGH THE CONSTELLATIONS,
1995-1998.

"down under," and by about September we in the northern hemisphere may also be able to glimpse it low above the southeastern horizon just before dawn. The comet may still possibly be bright enough to be seen with the naked eye at this time, although more than likely we'll need a pair of binoculars to pick it up then. After about late October, though, we northerners will lose Hale-Bopp entirely; the comet will be continuing its southward plunge, and by late 1997 it will be deep in the southern sky, not too far from the sky's south pole. Here it will stay as it makes its way back out into the far reaches of our solar system. Amateur astronomers in the southern hemisphere should be able to follow it with their telescopes for another year or more, and the large observatories there may be able to follow it well into the next century. It will not reach the distance from the sun at which Halley's Comet was when it was last photographed, until late 2003. Eventually, we Earthlings will lose sight of Comet Hale-Bopp completely, at least until it comes back again sometime around the 55th Century.

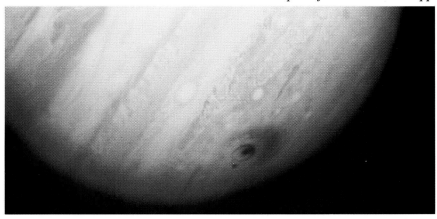

AFTERMATH. Jupiter's atmosphere after one of the impacts from Comet Shoemaker-Levy 9. The "scar," produced by material heated and churned up by the impact, is larger than the planet Earth. Image taken with the Hubble Space Telescope. *NASA photograph.*

COMET IKEYA-SEKI IN 1965. Photograph taken from Los Angeles by William Liller. Used with permission.

COMET WEST. A splendid object in the pre-dawn skies of March 1976. Photograph taken by Dennis di Cicco, used with permission.

THE RURAL NIGHTTIME SKY. The Milky Way arches gracefully through the summertime skies in rural New Mexico. Image provided courtesy of the International Dark-Sky Association.

HALLEY'S COMET IN 1986. Photograph taken by Tony and Daphne Hallas. Copyright Astro Photo.

HUBBLE SPACE TELESCOPE *IMAGE OF COMET HALE-BOPP, taken September 26, 1995. This image is in "false-color," used to enhance subtle details in the coma. NASA photograph.*

ARTIST'S CONCEPTION OF COMET HALE-BOPP over the southwestern American desert during the spring of 1997. Painting by Dan Durda, used with permission.

HALLEY'S COMET ON MARCH 7, 1986. Photograph by the author. This was a 30-second exposure on ASA 1000 film, utilizing the "camera-on-a-tripod" setup described on page 120.

TOTAL ECLIPSE OF THE MOON. Note the reddish color of the eclipsed moon, due to sunlight being refracted by the earth's atmosphere. Photograph by the author.

A BRIGHT COMET. The Daylight Comet of 1910, which appeared a few months before Halley's Comet that same year. Painting by Kim Poor, used with permission.

THE EFFECTS OF TRAILING. On March 27, 1996 Comet Hyakutake passed near the north pole of the sky. This 5-minute exposure by the author shows the north star Polaris (above the coma) and the comet's tail stretching back toward the Big Dipper. Note how the star images become more trailed the further they are from the pole.

A BRIGHT COMET – HALE-BOPP AT ITS NEXT RETURN? – AS SEEN FROM THE VALLES MARINERIS CANYON ON THE SURFACE ON MARS. Will representatives of humanity someday view this scene? Painting by Kim Poor, used with permission.

SPRING AND SUMMER 1996

During the summer months of 1996 Comet Hale-Bopp will be visible in the southern part of the sky, and relatively well placed for viewing. As I said above, the comet will be near its opposition at that time, rising in the southeast around sunset, appearing due south near midnight (which, remember, is really 1:00 AM for those whose locales "spring forward" to daylight savings time), and sets in the southwest near sunrise. How high the comet is at midnight depends on the viewer's latitude; a person located near a latitude of 40° north (e.g., Denver, Philadelphia) will see the comet almost halfway between the southern horizon and the zenith. Those south of this latitude will see the comet higher in the sky, those north of this will see it lower.

During most of July Comet Hale-Bopp will be seen crossing the rather obscure constellation of Scutum, the shield, which is located a little to the north of the prominent "teapot" shape of the constellation Sagittarius. The Milky Way's band is rather bright in Scutum, due to dense "clouds" of distant stars, and thus the comet may not appear quite as impressive as it would if it were located against a darker sky background.

During this time the comet continues to march slowly to the northwest against the background stars, and during the latter part of July it enters the constellation Serpens (the serpent) and in early August it enters Ophiuchus, the serpent-bearer (which supposedly depicts Æsculapius, the Roman god of medicine). By this time the earth has moved ahead of the comet's incoming path, and as a result the comet, along with these constellations, has moved into the evening sky. Since the comet is still somewhat far out in its orbit, its apparent path in the sky still exhibits the "back-and-forth" motion that is reflexive of the earth's motion in its orbit, and thus at the end of September the comet will appear to come almost to a stop relative to the background stars for a few days. Subsequently, with the earth now starting to pass over to the far side of the sun (as seen from the comet), the comet's apparent motion in the sky will start to turn toward the northeast.

As is true with every other phase of the comet's visibility, predictions as to how bright it will be during the summer are problematical. Based upon its brightness after it reappeared from behind the sun in February, it seems likely that for most of the summer an ordinary pair of binoculars should be sufficient for picking it up. One should look for a moderately large diffuse object – this being the coma – with a bright starlike object – the condensation – near the center. If someone suspects that he/she does have the comet, but can't be sure, one should look at it again the next night, by which time it will have moved against the background stars. It's hard to predict at this point just how long the tail will be during this period, but more than likely it won't look too spectacular, especially during July. With the earth being almost directly between the comet and the sun then, the tail will be pointing almost directly away from us. Towards September, though, our aspect will start to improve, and we'll have a more "broadside" view of the tail. At that time we should see perhaps a degree or more of filmy tail material pointing toward the east.

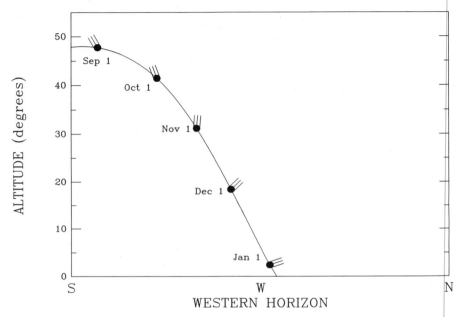

POSITION OF COMET HALE-BOPP ABOVE THE WESTERN HORIZON, SEPTEMBER-DECEMBER 1996. Unless otherwise specified, this and all subsequent horizon plots are for latitude 35° north, and show the comet's position above the horizon one hour after sunset (or before sunrise). Note the vertical scale varies from plot to plot.

Sometime during this period Comet Hale-Bopp could start to become visible to the naked eye, although just when this occurs can't be predicted with any accuracy. A lot of this depends upon the eyes and experience of the potential observer, how far he/she is away from city lights, and other factors unrelated to the comet itself. It's worth noting that the comet will be about 3½ AU from the sun in early August, and it is at this distance that the temperature in the comet's vicinity will start to become warm enough for ordinary water ice to sublimate. We don't know how much water Comet Hale-Bopp actually has; we do know that it has a lot of carbon monoxide, but for most comets it is water that produces the majority of eruptive activity which leads to the comet's being bright. If Hale-Bopp contains a water-to-carbon monoxide ratio that is typical of most comets, as soon as this water starts to sublimate we may witness a dramatic increase in its brightness within a relatively short period of time. Even if this does not happen, it is reasonable to expect that, by the end of September, Comet Hale-Bopp will start to become visible to the naked eyes of potential observers located in dark sites. Readers who are interested in performing this feat should make sure they are located far from city lights, with dark skies toward the western horizon, and may want to try locating the comet in binoculars first before trying to find it with the unaided eye.

One thing which potential comet observers should keep in mind is that a bright sky, whether it's caused by city lights or by moonlight, will make the job of viewing the comet much more difficult than it otherwise would be. While one can always (theoretically, at least) escape the city lights, one can't do too much about moonlight. Near the times of the full moon the entire sky is brightened for the duration of the night, and engaging in observations becomes much more problematical. (As I mentioned in Chapter 2, the full moons are when most astronomers catch up on their sleep.) I would thus recommend that readers avoid the times near the full moon when trying to find Comet Hale-Bopp. During the summer of 1996 full moons occur on the nights of June 1, June 30, July 29, August 28, and September 26, and any observations within two or three days of those dates probably aren't worth attempting.

There is one hope, incidentally, of observing the comet during the September full moon, and of picking up a rather unique observation at that. On the evening of the 26th not only is the moon full, but it dips into the earth's shadow, causing a total lunar eclipse. For the 70 or so minutes that the moon is immersed within the shadow, it will take on a reddish color (due to the sunlight being bent, or *refracted*, by the earth's atmosphere), but in the meantime the dimmer stars and other objects which would otherwise be obscured by the bright moonlight will make appearances in the temporarily darkened sky. The eclipse occurs during the mid-evening hours (totality lasting from 8:19 PM to 9:29 PM Mountain Daylight Time), and Comet Hale-Bopp, then located in the evening sky, should be easily visible to anyone observing from a dark site.

Speaking of the moon, one interesting event will occur on May 8, unfortunately before this book will reach its readers. On that morning the moon – then a couple of days shy of its third quarter phase – will *occult*, or pass directly in front of, the comet. In addition to being dramatic events in and of themselves, occultations – as these events are called – are quite useful from a scientific point of view, and have been used to collect information such as size, structure, etc., of the object being occulted (usually a star or a planet). This seems to be the first time an occultation of a comet by the moon will be observed, and at least some scientists have expressed the hope that this event will help in determining the size of the comet's nucleus and will aid in seeing how the various materials are distributed throughout the coma. The occultation will be visible from most of the western United States, with the moon and comet being located high in the southeast during the early morning hours (from approximately 4:00 AM to 5:00 AM, Mountain Daylight Time).

OCTOBER–DECEMBER 1996

During the last three months of 1996 Comet Hale-Bopp will be visible in the evening sky, and should be continuing to brighten gradually as the months progress. After late September the comet will begin traveling northeast against the background stars, accelerating its rate of motion as it does so, although due to the earth's motion around the sun the comet's position will sink lower and lower toward the western horizon.

In early October the comet is still fairly high up in the west after the sky becomes dark, and with the moon then being in the morning sky comet watching should be fairly easy. Many readers, in fact, may find this particular period to be the best time to view Hale-Bopp during the latter part of 1996. By this time it should be visible to the naked eye from any relatively dark viewing site, with binoculars or a small telescope revealing a few degrees of tail extending toward the northeast. The comet during this period is located in a rather "blank" part of the constellation Ophiuchus, i.e., there aren't any bright stars around to guide someone to the comet, and thus it may take a few minutes for someone to track it down in the sky (I'd recommend binoculars for this.) On the other hand, there aren't any bright stars to get in the way.

The moon will be back in the evening sky during the latter part of October, and is full on the 26th. Afterwards the comet will again be visible in a dark sky, and will be located somewhat to the northeast (relative to the background stars) of where it was found a month earlier, although by this time it will be lower in the western sky. While it should be brighter than it was earlier – perhaps even to the point at which it can be found with the naked eye without having to use binoculars to track it down first – the tail may not appear quite as long, since the earth's orbital motion will be causing us to start looking "down" the tail. It is possible, of course, that the tail may grow longer in physical terms during the interim, and this may act to compensate for our varying perspective.

Once again, the moon will interfere with comet observing during the latter part of November, with full moon occurring on the night of the 24th. By Thanksgiving we'll again have a dark sky for comet viewing, and by this time Hale-Bopp should be bright enough that prospective observers – especially those who have been following it all along – should have no trouble finding it. It will be quite low in the western sky, though, and those interested in looking at it should make sure they have a clear, unobstructed western horizon and should start looking for it almost due west during late dusk. Throughout early December the comet will sink lower into the western sky with each passing evening, although this may be partially offset by its increasing brightness (assuming, as always, that the comet brightens "as it should"). After about mid-December observing will become more difficult as the moon makes its way back into the evening sky; it will be full on Christmas Eve.

Although we'll have dark skies in the evening during the last week of 1996, by this time the comet will be very low in the western sky, and will set by the end of dusk. Nevertheless, by this time it may well be as bright, or brighter, than any of the stars in its vicinity, and patient observers with clear western horizons should still be able to pick it out of the early evening twilight. The further north one is located, the easier this task will be, although the task still becomes more difficult with each successive night. By year's end the comet will set by about the middle of dusk for most observers in mid-northern latitudes, and unless it happens to be quite a bit brighter than predicted at that time, viewing it will probably be rather difficult.

JANUARY–MARCH 1997

On New Year's Eve Comet Hale-Bopp is located directly north of the sun. (In astronomer-speak, we would say that the comet is in *conjunction* with the sun.) Theoretically, at that time the comet should be visible low in the west after sunset, and again low in the east before sunrise. Practically speaking, though, this observation will be very difficult, and thus for a few days most of us in mid-northern latitudes will probably find the comet to be invisible. (Observers at more northerly latitudes, say, the northern U.S., Canada, and northern Europe and Asia, may succeed in following the comet through conjunction, that is, if they can stand the cold temperatures.) After about the first week of January, though, the comet will begin to be visible in the morning sky, at first being rather low in the dawn, but rising noticeably higher in the sky with each passing morning. By this time the comet should be easy to find with the naked eye, perhaps being as bright as the stars in the Big Dipper (which will be heading down in the west on those chilly January mornings). The biggest down side at this time will be the fact that the comet's tail will be aimed almost directly away from the earth, and thus the tail we see will be greatly foreshortened. On the other hand, the geometry for viewing the tail will improve on a daily basis, and it might be an interesting project to see how the tail's apparent length changes during these days as a result of this.

Incidentally, as I mentioned toward the end of Chapter 2, there is a potential for a meteor shower from Comet Hale-Bopp on or around January 3. The sky location from which any meteors would appear to emanate will be fairly high up in the east at that time, and although there will be some moonlight to contend with – the moon's third quarter phase having occurred a couple of days earlier – this shouldn't be much of a problem. Don't confuse any Hale-Bopp meteors with the Quadrantid shower, which should peak about the same morning; the Quadrantids will be appearing to emanate from a spot higher in the north than those coming from the Hale-Bopp orbit.

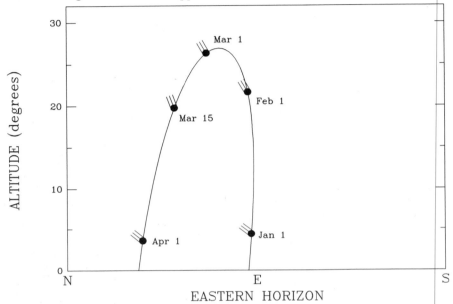

POSITION OF COMET HALE-BOPP ABOVE THE EASTERN HORIZON, JANUARY-MARCH 1997.

As January progresses the comet's visibility and appearance should improve noticeably with each passing day, becoming brighter, exhibiting a longer tail, and being visible in a darker sky. It continues to move northeast against the background stars, now perhaps rather obviously from one morning to the next. During the third week of January Hale-Bopp will be located about 10° (about the height of a fist held at arm's length) due west of the bright star Altair in the constellation Aquila, and may possibly be about as bright as that star. (Don't be too surprised, though, to find it somewhat dimmer – or, possibly, somewhat brighter – than Altair.) The moon, which is full on the 23rd, will interfere with observing during the latter days of January, although by this time the comet may well be bright enough that astronomers like me will probably forgo their monthly rations of sleep and keep on observing the comet anyway.

By early February the morning sky should once again be fairly dark, and if all goes well Comet Hale-Bopp should really begin picking up steam around this time. It continues its northeastward march against the stars, paralleling the main beam of the "northern cross" figure of the constellation Cygnus located a few degrees to its northwest. By mid-month it could easily be one of the brightest objects, if not *the* brightest object, in the predawn sky, glistening in the frosty air well above the northeastern horizon in the hours before twilight. Although this can't be predicted with any certainty, a rather substantial tail may be accompanying the coma in its travels. After the full moon on the 22nd the sight may be diminished somewhat for a few days because of the bright background sky, but by the end of the month the sky should once again be dark enough to allow the comet's splendor to show through. It is at this time that the comet will be at its highest elevation above the horizon in the predawn sky; during the following weeks it will begin to sink lower in the northeast.

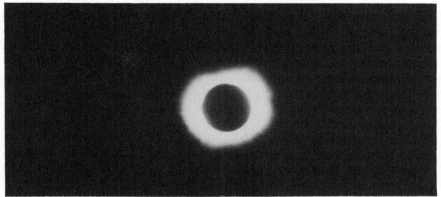

TOTAL ECLIPSE OF THE SUN. Note the corona, the sun's outer atmosphere. Photograph by the author.

On March 4 Comet Hale-Bopp will be located almost directly "above" the sun, some 1.04 AU (97 million miles) from it. From that vantage point, any riders on the comet will be able to look "down" at the solar system and would see, in theory at least, the planets orbiting in almost-perfect circles around the central "bull's-eye" of the sun. From our vantage point on the earth we'll see the comet some 15° southeast of the bright star Deneb at the northern tip of the Cygnus (the "head" of the northern cross), and hopefully quite a bit brighter than that celestial luminary. It continues on its rapid northeast trek against the stars, and by this time it is moving fast enough that its day-to-day motion should be easily discernible to a careful viewer.

A most unusual event occurs on Sunday, March 9: a total eclipse of the sun will be visible across sections of northern Mongolia and eastern Siberia. These events which, on the average, occur about once a year, are due to special alignments between the sun, moon, and Earth, and to many people – myself included – rank among the most spectacular sights presented to us by the heavens. For a few brief moments along a narrow track on the earth's surface, day is turned to night, the sun's faint (but hot) outer atmosphere, called the *corona,* makes its appearance, and many of the brighter stars and planets become visible. During the past two to three decades the sport of "eclipse-chasing," that is, of traveling to the various corners of the world in order to witness those few minutes of darkness, has become a rather popular pastime among some astronomers as well as many other folks.

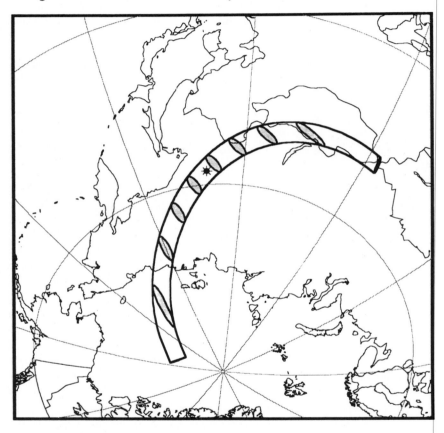

PATH OF TOTALITY FOR THE MARCH 9, 1997 TOTAL SOLAR ECLIPSE. From NASA Reference Publication 1369, courtesy Fred Espenak.

Ordinarily, the March 1997 eclipse would probably not be considered a high priority among eclipse-chasers; Siberian winters are notorious, of course, for their intense cold, and this particular part of the world would only be beginning to emerge from winter's grip. But as it turns out this eclipse will offer a special treat, in that during the at-most 2¾ minutes of totality Comet Hale-Bopp should be easily visible – perhaps even brilliant – in the darkened sky, some 45° – about ¼ of the sky – north of the eclipsed sun. Such eclipse comets – especially those bright enough to be seen with the naked eye – are extremely rare events, and only a handful have been observed throughout recorded history (the most recent substantiated occurrence being in 1948). Eclipse-chasers willing to brave the cold may thus be rewarded with the spectacle of two of the most beautiful celestial phenomena being visible simultaneously.

Even those viewers who decide to stay at home should still find Comet Hale-Bopp a spectacular object in the predawn sky during early March, still fairly well up above the northeastern horizon by the time the glow of dawn begins. By this time the comet should be brighter by far than any of the stars that might be around, and with the possible exception of Jupiter, which will be beginning to be visible low in the southeast, should outshine just about every object in the morning sky.

As March progresses the comet will continue sinking lower into the northeastern sky, although this should be partially offset by its continued brightening. Viewers at more northerly latitudes may start to notice that, in addition to being visible in the morning before dawn, the comet can also be seen low in the northwest after sunset. By the last week or two of March most viewers in the United States will be able to witness this phenomenon, and for those who are north of latitude 45° – i.e., the northernmost U.S., plus all of Canada and much of northern Europe – the comet will not even set, but will remain above the horizon all night, skirting the northern horizon during the hours around midnight. The comet, in fact, is in conjunction with the sun on March 22 – incidentally, the same day it is closest to the earth – but because it is located so far north of the sun at that time this unusual type of visibility becomes possible. After this date the comet should still be visible in the morning sky – and the further north one is, the longer this will last – but will start to make a better appearance in the evening sky after sunset.

Full moon occurs on the evening of the 23rd, and while ordinarily this would tend to cut the comet's visibility somewhat, this time some of us get a break: a partial eclipse of the moon occurs then. (Eclipses of the moon occur when it dips into the earth's shadow, but because it doesn't go all the way into the central shadow this time, this is why we see it as a "partial.") This partial eclipse is deep enough so that much of the moon's light will be effectively dimmed, and thus for a couple of hours the comet's glow should brighten up the night sky relatively unhindered. The middle of the eclipse occurs at 9:40 PM Mountain Standard Time and thus the western part of the United States and Canada will benefit the most from this event. Further east, the eclipse will occur later in the evening, and thus the comet will be quite a bit lower, if not already having set; those above latitude 45° will still have the comet above the horizon, though.

After the full moon the comet should be a brilliant object in the northwest during and immediately after dusk, and gaining in altitude each evening. On the evening of the 25th it is located just 5° north of the famous Andromeda Galaxy, and those with a bent toward astro-photography might try to record the two objects on the same piece of film. (The comet, of course, should be *much* brighter than the galaxy.) This is also as far north as the comet gets, with respect to the background stars; after this, its path begins to turn toward the southeast.

As good as the comet's display has (hopefully) been so far, we have reason to believe that everything seen up to this point will have been little more than an opening act. As March closes Comet Hale-Bopp is making its final approach to perihelion, and based on what we've seen from comets in the past, only now would the stage be set for the comet's grandest display.

APRIL–JUNE 1997

Comet Hale-Bopp makes its closest approach to the sun during the early hours on April 1, and – assuming that nature isn't playing some kind of April Fool's joke on us – we should be in for a treat during the following few weeks. Although this isn't true for all comets, a majority of them are brighter after perihelion than before; this is probably due to the sun's heat – which is greatest at perihelion – spurring on more and more eruptions of gas and dust. This is also when some comets split into pieces, introducing more material to be heated, and leading to dramatic surges in brightness. Normally, we would think that Hale-Bopp doesn't get close enough to the sun for the chances of this occurring to be too great, but it isn't unprecedented, so such an outburst is possible. Usually, the tail is longer after perihelion – sometimes significantly so – as it is only now that the material ejected at perihelion can be swept backwards into the tail. The fact that Hale-Bopp is conveniently visible in the evening sky now – as opposed to making any potential viewers arise in the morning before dawn – combined with the warmer temperatures associated with springtime, should enable it to be enjoyed by just about any interested viewer in the northern hemisphere with a decently clear sky. It is very possible that Comet Hale-Bopp will be seen by more people than any other comet in history.

During those early April evenings after perihelion the comet should be a glorious sight in the evening sky, glistening in the northwest with a brightness perhaps well in excess of any other celestial object that's visible. The moon will be out of the evening sky at the time, and thus anyone who is able to get away from city lights and can have a truly dark sky should be able to witness a truly awe-inspiring spectacle. If the show is anything like that exhibited by Comet West in 1976 – and all early indications are that Hale-Bopp should be at least roughly comparable – it should be a sight that no one will forget. If the comet develops a respectable tail – say, half an AU or more in length – our view of it in early April should be fairly "broadside," and it may extend for a significant fraction of the sky's expanse, extending upward from the coma.

I should mention here the prospects of the comet's being visible during the daytime. It may come as a surprise to learn that some of the brighter planets and stars can be visible during the day through a telescope, if care is taken in aiming it in the right direction. The planet Venus – normally, the brightest object in the sky apart from the sun and the moon – is actually an easy object to observe in the daytime sky, and can usually be seen with the naked eye without difficulty if one knows where to look. (My experience has been that Venus can be difficult to locate in the daytime sky, but once found I can usually see it without any problems.) It happens on rare occasions – perhaps two or three times per century – that a comet will become bright enough to be visible in daylight; the great sungrazing comet of 1882 and Comet Ikeya-Seki in 1965 were both so bright that all one had to do was block the sun with one's hand (or hide it behind a building) and the comet, including a decent tail, was readily apparent to the naked eye.

Will Comet Hale-Bopp become bright enough for daylight visibility? This cannot be answered with any certainty at this time, but if the comet brightens more or less as expected, it will probably *not* become bright enough for this to occur – at least, not with the naked eye or with binoculars, although experienced astronomers with large telescopes might succeed in picking it up. If, however, the comet exceeds the current expectations – and it wouldn't have to exceed them by too much – visibility during daytime is a distinct possibility. Finding it in the daylight sky might present some difficulties, though; it would help if the moon or Venus were close to it to aid in finding it, but neither of these objects will be near the comet during the times in question. One's best bet might be to notice where the comet is located with respect to a foreground object, such as a tree or building, during an evening, and then, while observing from the same spot the next afternoon, use that object as a reference point and then scan upward with a pair of binoculars. If the comet does seem bright enough on those early April evenings to warrant trying to see it in daytime, I would encourage readers to make this attempt; the sight of a bright comet in the daylight sky would certainly rank among the more unique astronomical sightings one could make.

Throughout most of early April the comet will be well-placed in the evening sky, high in the northwest. While it tracks to the southeast against the background stars during this time – this being noticeable as a southward shift relative to foreground objects from night to night – its altitude above the horizon won't change very much, and thus from mid-northern latitudes it is visible for a couple of hours after the end of dusk before setting. (Curiously, on the evening of the 9th it will be located at almost exactly the same spot in the sky that Comet Hyakutake occupied exactly one year earlier.) By about the middle of the second week of April a crescent moon will become visible in the evening sky – well to the south of the comet – and after about mid-month it will brighten the sky enough so that observations of the comet will begin to be affected. In particular, the apparent length of the tail will probably shorten quite a bit, since its outer portions will be washed out in the brightened sky.

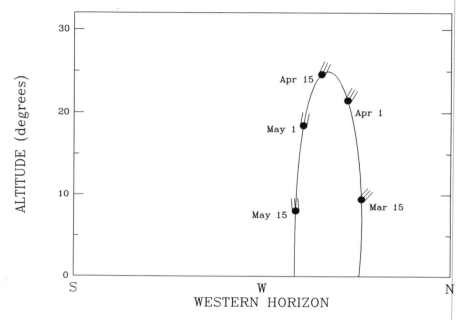

POSITION OF COMET HALE-BOPP ABOVE THE WESTERN HORIZON, APRIL-MAY 1997.

Full moon occurs on the 22nd, and within a couple of days the moon will have moved into the morning sky, once again leaving the evening sky dark for the comet. Now located well to the south of where it was earlier in the month, it is starting to sink lower in the west, and observers will notice that it doesn't remain above the horizon for as long after dusk. Also, the comet will likely be starting to fade by this time, although just how fast this proceeds is something that really can't be accurately predicted. One other thing which might detract from its visibility is that our view of the tail will become more and more foreshortened, so that even if – as seems possible – the tail becomes longer in physical terms, the apparent tail we see in the sky may well become shorter. Nevertheless, despite all these detracting factors Comet Hale-Bopp should still be a splendid object during the late April and early May evenings. It is during this period, in fact, that the northern hemisphere gets its last good look at the comet – for about 3400 years, anyway – and I'd recommend that comet watchers take advantage of this opportunity.

On May 6 the comet will cross the plane of the earth's orbit from "above" (i.e., north) to "below" (south), this occurring a little over 0.1 AU beyond the earth's orbit itself. At that time the comet is located a few degrees northeast of the Hyades star cluster in the constellation Taurus, the bull. (The Hyades form the distinct "V" shape that usually represents the bull's head.) From that time on the comet begins to sink more rapidly toward the western horizon, thus becoming more and more difficult to see well with each passing evening. By mid-month moonlight begins to interfere again, up until full moon on the evening of the 21st. Afterward the sky is dark again, but by this time the comet will be very low in the west, setting during mid-dusk. Observers located in the more southerly latitudes, such as my own location at 33°, will fare slightly better than those further north, but even from here by late May the comet will be setting only an hour or so after sunset. How long we're able to follow the comet depends to a fairly large extent on how bright it is at this time, but even under the more optimistic scenarios it would seem likely that we in the mid-northern latitudes will lose the comet somewhere around the end of May.

The situation is a bit different for our friends located in the southern hemisphere, such as South America, southern Africa, Australia and New Zealand. Up until now these folks have been left out, having lost sight of the comet back about November 1996. At the time of its perihelion passage Hale-Bopp is still not accessible to southern hemisphere viewers, but by about the end of April they should be able to catch sight of it along their northwestern horizon during the early evenings. As April progresses into May the comet rises higher into their evening sky, until by late May, when we're losing it, they'll still have it fairly high above the horizon well after the end of dusk. Although the comet will probably have faded somewhat from its peak brightness by this time, it is possible that a fairly long tail will be visible at this time – such as what happened with the Great Comet of 1811 – and thus it is rather possible that our southern hemisphere friends will have a decent display to admire after all.

Still continuing to track to the southeast against the background stars, on June 5 Hale-Bopp will be situated just a couple of degrees northeast of the bright star Betelguese located at the northeast corner of the constellation Orion; the two objects may be of similar brightness. After about this time the comet will begin sinking closer to the horizon even as seen from the southern hemisphere, although this is a gradual process, and it should remain conspicuous throughout the remainder of June. The moon, at its first quarter phase on the 13th, will start to interfere with viewing around then, and will continue doing so until full moon, which occurs on the 20th.

A rather interesting event will occur around mid-month, which will be affected slightly (although presumably not drastically) by the moonlight.

Encke's Comet, the comet with the shortest known period of 3.3 years (see Chapter 1) will also be visible in the evening sky from the southern hemisphere during June. (It passes its perihelion on May 23, 0.33 AU from the sun.) On the 15th it will be located only 3½° east of Comet Hale-Bopp, and depending upon the length of the latter object's tail and how much curvature it is exhibiting, at some time around that date Encke's Comet will be situated directly in front of Hale-Bopp's tail. At that time Encke's Comet is 0.42 AU from the earth, compared to 2.46 AU for Hale-Bopp. Observations from previous returns suggest that Encke's Comet should be bright enough to be easily seen in binoculars at this time, and may even be close to naked-eye visibility. Would-be astro-photographers might want to try recording this interesting and unusual meeting of comets. (Incidentally, Encke's Comet will pass 0.19 AU from the earth in early July, the closest approach it has made since its original discovery over two centuries ago.)

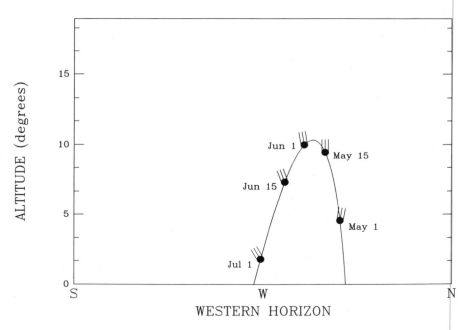

POSITION OF COMET HALE-BOPP ABOVE THE WESTERN HORIZON, APRIL-JUNE 1997, AS SEEN FROM THE SOUTHERN HEMISPHERE. Plot is for latitude 35° south.

After the full moon Comet Hale-Bopp should again be visible in the southern hemisphere's evening sky, although by this time it is beginning to sink distinctly closer to the horizon, and it will also probably be starting to fade significantly. Nevertheless, it should still be visible to the naked eye from a dark site without any difficulty. By the end of June it will set before the end of dusk, and viewers located at the more southerly latitudes – say, 40° or more south – may also begin to see the comet rising in the morning during dawn. On July 3 the comet will be in conjunction with the sun for the third time in six months, and after that becomes a morning-sky object for the entire southern hemisphere.

MID-1997 AND BEYOND

After its perihelion passage on April 1 Comet Hale-Bopp begins a dramatic southward plunge, rapidly heading almost "straight down" through the orbital planes of the earth and the other planets. (When near aphelion some 1700 years from now it will be located over 300 AU "below" the planets' orbital planes.) As a result, viewing of the comet as it leaves the inner solar system will be almost – but not quite – the exclusive domain of comet watchers in the southern hemisphere.

Following its conjunction with the sun in early July the comet is visible in the morning sky from suitable locations "down under," trekking southeast through the rather obscure constellation of Monoceros, the unicorn (in what we northerners might think of as the "winter" Milky Way.) It may still be visible to the naked eye without much difficulty at this time, although very likely it won't exhibit the brilliance and splendor that it displayed around perihelion. Nevertheless, it may still possess a few degrees of tail, extending to the southwest from the hazy coma. In late July Hale-Bopp is located some 10° from the brilliant star Sirius – the so-called "Dog Star" – although bright moonlight in the morning sky then might detract a little from this meeting.

Over the next few months the comet continues its southward motion, traversing through the constellation Puppis (the stern of the ship *Argo*, from the mythological journey of the Greek hero Jason), and rising ever higher in the southern hemisphere's morning sky. At the same time it continues fading, and sometime during these months its period of naked-eye visibility will come to an end. This is due to several factors, the most important ones being the comet's ever-increasing distances from the sun and the earth. As its distance from the sun increases the comet's nucleus will start to become less and less active; i.e., there will be fewer eruptions of ice and dust from its surface, and those that do occur will be of lesser intensity. At some point – and there's no real way to say when, since this varies from comet to comet – the nucleus will shut down almost entirely. All that remains then is a coma that is no longer being replenished, and over a period of weeks and months the material in the coma gradually disperses into space. From our perspective, what we see is the coma gradually growing larger, dimmer, and more diffuse, until finally it "diffuses out" altogether. Based upon Hale-Bopp's activity at a large distance as it approached the sun, this point may not occur on the outward trek until quite some time after perihelion, but even so comet watchers will probably notice the coma as getting larger and hazier as the comet recedes. Naked-eye visibility may cease, then, not so much from the comet's overall brightness getting dimmer, but simply because the coma eventually will just expand out until it is just too dim to see. Most likely this will occur sometime between August and October, although this could easily be off by a month or more in either direction.

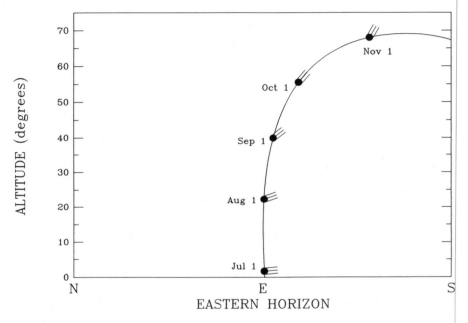

POSITION OF COMET HALE-BOPP ABOVE THE EASTERN HORIZON, JULY-DECEMBER 1997, AS SEEN FROM THE SOUTHERN HEMISPHERE. Plot is for latitude 35° south.

Despite the comet's southward flight, comet watchers in mid-northern latitudes should get one final shot at it during the early autumn. In early September Hale-Bopp will rise around the beginning of dawn, appearing very low above the southeastern horizon before the brightening sky wipes it out. The situation improves a little with each morning, although moonlight will interfere some during the latter part of the month. (Full moon occurs on the 16th, and our friends in the eastern hemisphere will be treated to a total eclipse of the moon on that date. Folks in southeastern Asia – i.e., in China and Japan – may succeed in picking up the comet during the eclipse.)

The best opportunities for viewing Hale-Bopp then will come in early October, when the moon is out of the way, and the comet now rises long enough before dawn to be visible in a reasonably dark sky. Nevertheless, it will remain fairly low above the horizon, never rising high enough to get out of the "murk" that interferes with viewing objects at low elevation. The further south one is located, the better the visibility will be, although even those at my latitude will not get to see much. (Those near the northern border of the U.S., and points further north, probably won't get to see it at all.) It's difficult to predict just how bright the comet will be at this point; more than likely, it won't be visible to the naked eye, but a decent pair of binoculars should be sufficient for picking up the coma's hazy glow. In other words, Comet Hale-Bopp probably won't be too much to look at then, but since this is the last opportunity most of us will get to see it, it's an opportunity most comet watchers will probably want to take advantage of.

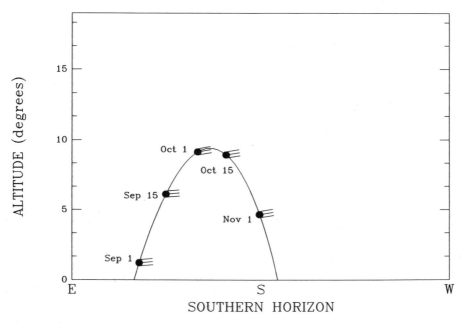

POSITION OF COMET HALE-BOPP ABOVE THE SOUTHERN HORIZON, SEPTEMBER-NOVEMBER 1997, AS SEEN FROM THE NORTHERN HEMISPHERE. Plot is for latitude 35° north.

The moon is full on October 16, and thus the bright sky will interfere with comet viewing for a while after that date. By the time we have a dark sky again – around the end of the month – the comet will be quite a bit further south than it was earlier in the month, and although theoretically visible for viewers near my latitude, it will remain quite close to the southern horizon. It is possible that those located in the southernmost parts of the continental U.S. and who have a clear horizon to the south may be able to view the comet during the first couple of weeks of November, but even this will cease after about mid-month, because of both the comet's continuing southward motion and the full moon that occurs on the 14th. After that, except for those who are located in tropical latitudes, Comet Hale-Bopp is gone as far as the northern hemisphere is concerned.

After about mid-November those in mid-southern latitudes – for example, central Australia – will find that the comet is far enough south so that it is above the horizon all night, never setting. As it turns out, this state of affairs will remain from that point on, giving our southern hemisphere friends the opportunity to keep a more-or-less continuous eye on Hale-Bopp as it recedes into the outer solar system. At the end of 1997 the comet is very well placed for observation from the southern hemisphere, being near opposition and thus highest around midnight, and is located a few degrees northeast of the Large Magellanic Cloud (a satellite galaxy of the Milky Way, which gained popular fame in 1987 when a supernova erupted within it). Over the next few months the comet makes a slow loop over the northernmost extremities of this object. It may still be visible in binoculars during these months, although most southern comet watchers will probably need a small telescope to follow it well.

Although by this time the comet is long gone as far as we in the U.S. are concerned, anyone who journeys to tropical latitudes in the northern hemisphere still has a chance to see it. A good excuse for doing so occurs on Thursday, February 26, 1998, when the path of another total solar eclipse crosses northern South America and then across the central Caribbean Sea. The southern Caribbean islands of Aruba and Curaçao are both in the path of totality, as are the eastern islands of Montserrat, Antigua and Guadeloupe. Eclipse chasers from the U.S. and other mid-northern regions will also have the opportunity to explore parts of the nighttime sky not normally accessible to them, and among these will be the obscure southern constellation of Dorado (the swordfish), wherein will lie Comet Hale-Bopp. From Guadeloupe (latitude 16° north), the comet will be visible low in the southwest after dusk, and probably will still be bright enough so that it can be seen in a relatively small telescope. Both Tom Bopp and I are planning to travel to the Caribbean to see this eclipse, and unless one or both of us makes a trip to the southern hemisphere afterward – which, at least in my case, is possible but unlikely – this will be the last time that either of us sees the comet. Thus, this event gives us the opportunity to, together, say good-by to this celestial object which will jointly bear our names throughout the rest of history.

After that, my southern hemisphere colleagues will have the comet to themselves as it fades away, making smaller and smaller loops (due to the

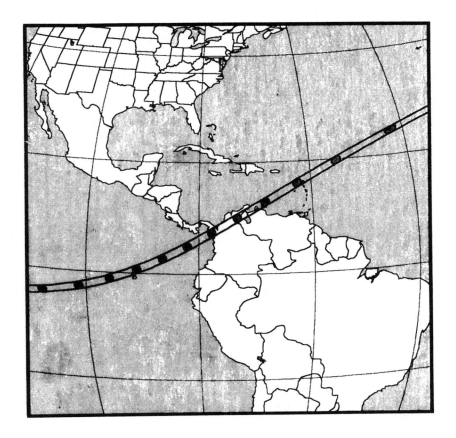

PATH OF TOTALITY FOR THE FEBRUARY 26, 1998 TOTAL SOLAR ECLIPSE. From NASA Reference Publication 1383, Courtesy Fred Espenak.

earth's motion around the sun) deeper and deeper in the southern sky during the following months and years. It should still be visible in a fairly small telescope when, in early July 1998, it passes just over a degree – twice the moon's diameter – northeast of the bright star Canopus (the second brightest star in the entire sky, after Sirius, and from my latitude visible low in the southern heavens during the winter months). In early December 1998 its distance from the sun will be 7.15 AU – the same distance it was the night Tom Bopp and I discovered it – and it will be located a few degrees southwest of the Large Magellanic Cloud. Depending upon how long its nucleus continues to erupt off gas and dust, amateur astronomers below the equator may still be able to pick it up with decent telescopes at that time.

117

Having followed comets myself for over 25 years, I suspect that my colleagues in Australia and elsewhere in the southern hemisphere will try to follow Comet Hale-Bopp for as long as possible. My guess is that the decently-equipped amateur astronomers will eventually lose sight of it sometime during the first half of 1999. Professional astronomers with large observatory telescopes – and perhaps a handful of amateur astronomers with access to modern imaging equipment – should still be able to follow it for some time after that. In mid-December 1999 the comet will be 10 AU from the sun – about the same distance as Saturn – and in March 2001 it will be 13.1 AU from the sun, the same distance as it was during Rob McNaught's photograph in April 1993. (The professional astronomers may enjoy following the comet during mid-2000, when it will track across the densest portion of the Large Magellanic Cloud, passing almost directly over the location of the 1987 supernova on June 9.) Near the end of 2003 it will be 18.8 AU from the sun, the same distance Halley's Comet was when it was last photographed in 1994, and almost as far away as the planet Uranus. The *Hubble Space Telescope*, if it is still operating, and/or any successor instruments, may continue to follow it on an occasional basis after that, if they can, and I suspect that it may not be until the latter part of the first decade of the 21st Century that the final images of the dim receding comet will be obtained. Then, and only then, can we justifiably say that Comet Hale-Bopp is truly "gone."

Incidentally, from about the middle of 2004 on until the time we finally lose it, Hale-Bopp stays within 10° of the south celestial pole. Not until the waning decades of the 21st Century, at which time it is over 100 AU from the sun, does it begin to inch its way further north. This is only of academic interest, of course, since it will be well beyond the reach of any telescope equipment we now have or expect to have anytime soon, but it's an interesting thought to consider that Hale-Bopp will remain in this one tiny region of the sky until long after Halley's Comet has come and gone again (the next return being in 2061).

EQUIPMENT YOU MAY NEED, AND DON'T NEED, AND OTHER TIPS

If all one wants to do is look at Comet Hale-Bopp, one really shouldn't need any equipment at all, just one's eyes (at least, during the early months of 1997 when the comet is bright). If, as we hope, the comet develops a long and impressive tail, the unaided eye is actually the best instrument one can use for seeing it; the restricted fields of view inherent in binoculars and telescopes will not be able to capture the entire spectacle in the way that one's eyes can. I thus recommend that those readers with a casual interest in the object think twice before purchasing any additional equipment.

Keeping in mind what I just said, a *good* pair of binoculars will certainly do a lot toward enhancing one's views of the comet. In addition to helping to accentuate details in the coma and the tail, binoculars can also be enormously helpful in locating the comet while it is still rather dim; i.e., during the latter months of 1996. (And, although I hate to bring it up, if the comet for some reason doesn't live up to the expectations we have for it, binoculars will certainly be useful in viewing it. But let's hope otherwise.) Normally, a decent small pair of binoculars, such as 7x35, 7x50, or 10x50 should be more than sufficient. ("10x50" means that the binoculars have a magnification of 10x, and have lenses 50 mm in diameter.) Larger binoculars do exist, but in addition to being more expensive, these have a tendency to be too heavy to hold steadily in one's hands, and usually require a tripod of some kind for steady support. I recommend that one be somewhat cautious in purchasing a pair of binoculars; I've included some reputable manufacturers in Appendix A for readers who might be interested in pursuing this.

For those who might be interested in acquiring a telescope for viewing Comet Hale-Bopp, I have a few cautionary words. First, a good telescope is not cheap, and such an investment probably shouldn't be made just for the comet alone; remember that the comet will not be around forever, and one probably doesn't want to be stuck with an investment of this nature that is no longer being used. There are, of course, plenty of other objects in the sky that can be looked at and enjoyed through a telescope, and if through following Hale-Bopp readers come to appreciate the sights of these objects, then perhaps acquiring a telescope isn't such a bad idea.

I suspect that many amateur astronomical clubs and other organizations throughout the world will be staging public viewing nights of the comet when it is bright. If no such group is near you, I am almost sure that an amateur astronomer of some kind isn't too far away, and you might want to make contact with him or her. Attending one of these viewing nights will allow you to see the comet through a telescope without the actual expense and hassle of purchasing one. It will also allow you to get a first-hand look at some of the telescopes that are available should you be interested in making such a purchase.

Finally – and if you read nothing else about telescopes in this section, please read this – if you do decide to purchase a telescope, please *don't* purchase one from an ordinary department store or through a department store's catalogue. Normally, these are mass-produced items of inferior quality, and are usually overpriced to boot. (I remember during the Halley's Comet era seeing several of these devices in department stores that were priced three to four times what they were worth.) There are specialized companies that deal exclusively in manufacturing telescopes and other optical devices, and I've included some of the reputable ones in Appendix A; I *strongly* recommend that anyone interested in purchasing a telescope contact these. One other avenue that prospective telescope buyers might pursue is to check out the classified ads in magazines like *Sky & Telescope* and *Astronomy* (see Appendix A); it is often possible through these ads to pick up a decent used telescope for a reasonable price.

One other thing I should mention with regard to telescopes is that what matters in selecting a telescope is not its "magnification," but rather the size of its lens (for a refracting telescope) or mirror (for a reflecting telescope). The larger the lens or mirror, the more light it can gather, and the more objects one can see with it. With almost any decent telescope, one can change the magnification simply by changing the eyepiece (and these can be purchased separately), and in many cases one doesn't want a lot of magnification anyway — certainly not for an extended object like a comet. Among other things, increasing the magnification increases the effects of the atmospheric turbulence between you and the object you're looking at, so there is usually a practical limit as to how much magnification you can use on an object. Most department store telescopes are advertised by their magnification, not by the size of their lens or mirror; another indication that you're dealing with firms that don't know too much about telescopes (and assume that you don't either).

For those who want to try the more challenging task of taking photographs of the comet, there are at least three main ways to proceed. By far the easiest of these is to simply mount a camera on a tripod, and open the shutter for a time exposure. This usually requires a 35 mm Single Lens Reflex (SLR) camera, with the ability to take time exposures, and moderately fast film (I'd recommend ASA 400 or faster). What one does is simply open the lens up to its widest possible f-stop (normally, f/1.8 for a standard 50 mm lens), open the shutter, and expose away. Because the earth is rotating during the exposure, and the camera isn't turning to compensate, if one exposes too long the comet and stars in the photograph will start to "trail." Thus, there is a limit as to how long one can expose the photograph and still come up with an æsthetically pleasing result; depending somewhat upon how far north (or south) the object of interest is, this limit is usually about 45 seconds to 1 minute for a 50 mm lens. The use of a telephoto lens will result in a larger image on the photograph, but normally one can't open these lenses as far as one can a smaller lens, and the higher magnification also decreases the length of the practical exposure time, before trailing begins. The best bet is to experiment with a variety of lenses, exposure times, and films, and see which give the best result. Be prepared to produce a lot of photographs that aren't too pretty, but some of the combinations should give some rather nice results.

The other two methods require not only a telescope, but a telescope equipped with a "clock drive" (a device which turns the telescope to compensate for the rotation of the earth). One method is to mount the bare camera, with whatever lens you're using, onto the tube of the telescope "piggy-back" style; this allows you to use longer time exposures without having to worry about the effects of trailing. The other method is to mount the camera at the eyepiece of the telescope, and shoot the photograph directly through the telescope's field of view. Both of these methods can produce very pleasing results, but at the cost of increased complexity, not the least of which is the fact that one has to continuously ensure that the telescope is tracking the photographed object accurately. The precise details of these methods are beyond the scope of this book, but I've included some references in Appendix A for those who might be interested in pursuing this further.

One recent technological advancement which has revolutionized the art of astro-photography has been the development of "Charge Coupled Devices," or CCDs. These are electronic imaging systems which utilize a computer chip to convert the light they receive into electrical charges, and then store the resulting information as a computer-readable image that can be stored within a computer's memory devices (e.g., a hard drive or a floppy disk). Today many amateur astronomers, and almost all professional astronomers, utilize CCDs in lieu of photographic film for taking astronomical photographs. Their primary advantage is that significantly fainter objects, and more detail within those objects, can be photographed within a small fraction of the exposure time that photographic film requires. For example, a 5-minute exposure with a typical CCD will often record dimmer objects than a 2 or 3 hour exposure with film. On the other hand, there are some drawbacks to using CCDs; one of these is that the field of view, which is limited by the size of the computer chip that is used, is quite small. The biggest drawback is probably price: CCDs are not cheap – even a relatively small and simple one will cost two to three thousand dollars – and their usage also requires the purchase of auxiliary computer equipment and software. My suspicion is that most readers will probably not be interested in pursuing this route, but for those who might be I am including some appropriate references and information in Appendix A.

Whatever equipment one decides to use in studying the comet, there is one issue which I've addressed several times throughout this book and which I want to stress heavily here. Because of the primarily urban society which we live in these days, a significant majority of our population do not get to experience the dark nighttime skies that our ancestors just a couple of generations ago almost took for granted. In many ways, this so-called "light pollution" is entirely unnecessary, since many of the lights within urban areas direct light upward, where it is completely wasted, instead of at the ground where it is needed. The usage of proper shielding around these lights would eliminate a lot of this waste and, in addition to allowing city dwellers the opportunity to appreciate the nighttime sky more, would also cut down drastically on power costs. Those who might be interested in pursuing this goal may want to work with some of their local elected leaders to help bring it about, and for the benefit of these individuals I've taken the liberty of listing organizations which exist for this purpose in Appendix A. In any event, though, people who live within large metropolitan areas are going to have to escape to some of the outlying rural areas in order to fully appreciate the spectacle of Comet Hale-Bopp.

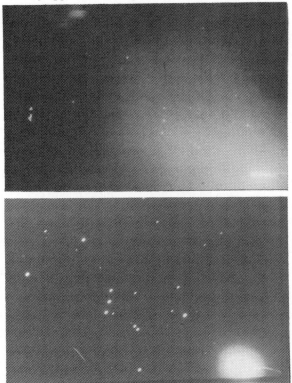

THE EFFECTS OF SHIELDING STREETLIGHTS. These scenes of the constellation Orion were taken from within a moderate-sized city. The scene at top shows the effects of unshielded streelights; the scene at bottom shows what happens when streetlights are properly shielded. Photographs provided courtesy of the International Dark-Sky Association.

WHAT WILL WE LEARN FROM COMET HALE-BOPP?

In the above descriptions about when and where to look for the comet I have been cautiously optimistic as to how bright and spectacular it might get. But, as I hope I've managed to convey throughout this entire discussion, we have no real way of knowing at this time if in fact the comet will behave like this. It is entirely possible that Comet Hale-Bopp will "fizzle" in a manner similar to that of Comet Kohoutek. Although it would seem unlikely, given its present brightness at such a large distance from the sun, there exists the possibility that

Hale-Bopp may not even become visible to the naked eye. If such a scenario were to happen, I and a great many other scientists and laypersons would be deeply disappointed, of course. Nevertheless, regardless of what happens, Hale-Bopp will certainly lead to significant advances in our understanding of comets and, by connection, the formation of the solar system and, ultimately, ourselves. It's worth pointing out here that, even though most of the general public considered Comet Kohoutek and the recent return of Halley's Comet as "duds," from a scientific point of view both of these objects were immensely satisfying.

Indeed, Comet Hale-Bopp is already generating an immense amount of interest among cometary astronomers; in the words of one such individual, Hale-Bopp "presents potentially the best opportunity in our lifetimes to study the changes in cometary properties with time and for unique types of remote sensing of a comet." Discussion has already started on a joint program between NASA and the National Science Foundation (NSF) for an intense scientific study of the comet over the next year and a half. A joint NASA-NSF steering committee organized to coordinate this effort has even recommended a series of "International Comet Hale-Bopp Days" – four specific periods wherein especially important scientific observations can be carried out. These include an interval in early May 1996 near the point where water ice might begin to sublimate; a period in September 1996 corresponding to the period during which Halley's Comet became active on its inbound approach to the sun; the period around perihelion and closest approach to the earth; and an interval in August and September 1997 when activity within the coma might begin to "shut down."

Already, Comet Hale-Bopp has been an enormously interesting object from the scientists' perspective. Never before has a comet been observed to be this bright and this active when so far from the sun. The comet has already given us quite a bit of information about the physical and chemical processes that can go on within comets at these kinds of distances and temperatures. Since Hale-Bopp is coming in quite a bit closer to the sun, we have before us an almost unprecedented opportunity to examine how these processes change over time and differences in temperature. Since Hale-Bopp seems like it may be a "large" comet in the physical sense, it's possible we may be able to detect the atoms and molecules of substances which are very rare in comets and which simply exist in too few numbers to be detectable in smaller comets. Regardless of whether or not detections of this kind are made, the substances we do see will be able to tell us a lot about the environment and processes that go on at distances of a few hundred AU from the sun. Since it is at these distances where many of the residuals left over from the formation of the solar system may reside, the study of Hale-Bopp may tell us quite a bit about the environment of the solar system at the time of its formation, and how it has changed during the time since then.

One more example of a potentially interesting observation we should be able to make with Hale-Bopp is the study of the solar wind. The structure of a comet's ionized gas tail can be significantly affected by variations in the solar wind, and since Hale-Bopp will traverse over a wide range of solar "latitudes," we should be able to detect these variations by seeing how its tail changes as it "sees" different areas of the sun. We've learned from the recently-completed European Space Agency (ESA) *Ulysses* mission that the solar wind does seem to change dramatically with solar latitude, and Hale-Bopp should be able to provide us with a new set of measurements of this phenomenon.

One reason why the study of Hale-Bopp should be so productive from a scientific standpoint is the current state of technology in telescopes and instrumentation, much of which didn't exist in its current state even as recently as the return of Halley's Comet in 1986. Just to cite one example, the *Hubble Space Telescope*, which was still on the ground during Halley's return, has already returned images and data from Hale-Bopp of a quality we've never seen before in a comet, and obviously this instrument will continue to provide us with unprecedented view of the comet as it develops. Ironically, *Hubble* is scheduled for its next servicing mission in February 1997 – in a schedule made out before Hale-Bopp's discovery – although this is certainly subject to change. At the same time, some of the advances in ground-based astronomical techniques have been almost as dramatic, and as these are loosed upon the comet during its appearance the amount of information we should receive should be simply incredible. Undoubtedly, Hale-Bopp should be the most thoroughly studied comet in history, and very likely will force substantial revisions in our textbooks by the time its appearance is over.

One question which is tempting to ask is whether or not any space probes will be sent to the comet during its appearance. Several such probes, most notably the European Space Agency's *Giotto* mission, flew by Halley's Comet during that object's return in 1986, and returned enormous amounts of data. There are some missions being planned right now, including ESA's *Rosetta* mission and NASA's *Stardust* probe, which will be launched within the next few years to rendezvous with a comet and – in *Stardust's* case – return samples to Earth, but these are scheduled to fly to known short-period comets in Jupiter's family. These objects have, by far, better-known orbits than other comets, and thus navigation is not as problematic as it would be for the longer-period objects. Also, most of the Jupiter-family comets have orbital planes fairly close to the plane of the Earth's orbit, and thus only a relatively small amount of fuel is necessary to get probes to them. Sending a probe out of the plane of the earth's orbit takes an enormous amount of energy, since we have to find some way to counter the velocity given to the probe due to the earth's motion around the sun. This either takes an enormous amount of fuel or, if we're willing to wait for a few years, we can take advantage of a large gravitational field, such as that of the planet Jupiter, to do the work for us. This latter method was used successfully by ESA's *Ulysses* spacecraft, which explored the high polar latitudes of the sun.

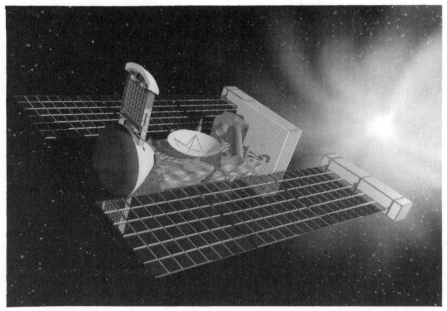

STARDUST. *This artist's conception shows the* Stardust *probe approaching the Comet Wild 2 in 2003. Image provided courtesy of Lockheed Martin Astronautics and Stephen Price.*

The idea of a probe to Hale-Bopp is complicated by the fact that the comet's orbit is almost perpendicular to the earth's orbit. The only practical times for a probe rendezvous with the comet would be on those two occasions when the comet intersects the earth's orbital plane. One of these has already occurred, in late February 1996, when Hale-Bopp went from "below" (i.e., south of) the earth's orbital plane to "above" (i.e., north of) it. This point is out close to Jupiter's orbit, and we didn't have enough advance warning to send a probe out that far. On the other hand, as mentioned earlier, Hale-Bopp will again cross the plane of the earth's orbit on May 6, 1997, at a point just over 0.1 AU beyond the earth's orbit itself. Theoretically, it is possible that a probe encounter could occur at about this time. There are, however, other factors to consider as well, such as the orbital speeds of the earth and of any such probe, and when these are factored into the equation, we find that there is no "easy" – which can probably be translated as "inexpensive" – way to get a probe to the right place at the right time. If any entity is willing to spend the fuel, it probably would be possible to

125

set up a probe rendezvous with the comet, but in the fiscally-tight climate under which NASA, ESA, and the world's other space agencies must currently operate, this is unlikely to take place. Also, it does take a certain amount of time to get a mission ready, even under the best of circumstances, and although the early discovery of Hale-Bopp does give us an unusually long time to prepare such a mission, it still probably isn't quite enough. Despite all this, however, it is nevertheless possible that such a mission might be sent after all; I have heard vague scuttlebutt to the effect that a mission of this nature is being informally discussed, and it is always possible that a private organization might try to launch its own probe.

In discussing what we hope to learn from Comet Hale-Bopp, we should include what the general public can learn from this event. In many ways, we scientists have a tendency to operate in our own little world, and to isolate ourselves from the public (other than to ask you to support our work through your tax dollars). I don't believe this is intentional; it's just that most of us are most comfortable when we're engaged in our scientific pursuits, and many of us don't have the time (or the ability) to communicate what we're doing to the public at large. The result has been the formation of gap between the public and the scientists, with growing distrust between the two groups. But as our society becomes more technologically oriented, and as our scientific research becomes ever more complex, I believe it is imperative that the public become as scientifically literate as possible in order to be able to make the necessary decisions as to the future direction of science.

Comet Hale-Bopp, in my opinion, presents an ideal means to help bring this about. Here is a unique phenomenon which, hopefully, will become bright enough so that anyone who wants to study it can do so without having to acquire a bunch of fancy equipment. Furthermore, it is something that most people should be able to experience directly, with their own eyes, rather than having to have it presented to them on television. By being able to follow the comet as it makes its visit through our skies, viewers should receive a dramatic introduction to some of the entities within our solar system, and to the processes by which they operate. By extension, this should help people understand how the various elements of the heavens "fit together," and give them an appreciation of the wonders that exist throughout the universe.

My primary purpose in writing this book has been to give the general public the background to understand and appreciate both Comet Hale-Bopp and the universe as a whole. In my discussions with other scientists I have encouraged them to use this opportunity to share with the public, in every way possible, this comet and whatever knowledge is learned from it. It is my hope that this celestial phenomenon can help bridge the gap between science and the public, and if this indeed were to occur, this may very well be Comet Hale-Bopp's most important and enduring legacy.

AFTERWORD: A PERSONAL VIEW OF COMET HALE-BOPP

Daddy, your comet is going to be in my science books one of these days, isn't it?
— Zachary Hale*

Little did I realize, when I walked outside the house late that Saturday evening in July 1995, what fate had in store for me for later that night. What happened during the course of that first hour I was outside has been nothing less than a life-changing event. The man who wheeled his telescope out of the garage onto the driveway was an all-but-unemployed professionally-trained scientist who was living primarily off his wife's salary and managing to earn a small income by writing space articles for a local newspaper, but in almost all other aspects was hitting a dead end in his professional career. In the course of that hour I went from being essentially a "nobody" to getting on the road to becoming one of the most sought-after scientists, if not individuals, on this planet. For the past several months I have been receiving numerous phone calls from media representatives from all over the world, am receiving requests for speaking engagements for numerous places, have received offers for agent services and employment opportunities, and have even managed to find a publisher willing to take on this book. This is only the beginning; if all goes well with the comet, by early next year I will be so completely inundated with requests for speaking engagements, tours, more book offers, and so on, that I won't be able to think straight. (I really hope I have some time to actually look at my comet!) My professional career and financial status both stand a reasonably good chance of being secure after all this is over, and I'll never be a "nobody" again. And all because I just happened to point my telescope at a particular point in the sky on that clear summer night.

To fully understand what this event in my life truly means to me, allow me to turn the clock back to my childhood days in Alamogordo, New Mexico. The skies in the desert southwest are usually quite clear, and with my family living on the southern outskirts of town, growing up with an interest in the sky was quite easy. I first became interested in astronomy as early as the First Grade, when my father checked out some books on the subject from the local library and handed them to me to look at. Like most elementary school kids, my interests oscillated between various subjects over the subsequent years, although by the time I was 12 years old I had settled on astronomy "for good." I have never lost my interest since then, and I never expect to.

During the time I was progressing through elementary school, our nation was caught up in the "rush-to-the-moon" frenzy that was initially spurred by the

---

* My older son, then 8 years old, over the breakfast table one morning.

SKY & TELESCOPE

NOVEMBER 1995 $3.95
$5.25 CAN.

Great Free Software
for Amateur Astronomers

The Puzzling X-ray Background

Ulugh Beg's 15th-Century
School for Astronomy

Could This Become The Next Great Comet?

Believed to be 10 times larger than
Halley's Comet, newly discovered
Comet Hale-Bopp may grow
100 times brighter.

SOME GOOD PUBLICITY. *The front cover of* Sky & Telescope, *November 1995 issue. Copyright 1995 by Sky Publishing Corporation, reprinted with permission.*

Soviet Union's launch of *Sputnik 1* in 1957. This state of affairs not only included the visible space program, but also carried with it an enormous emphasis on math and science education in the nation's schools. I was a beneficiary of this, and when I coupled my schooling with the nation's space exploits it was very easy for me to get excited about my own future. Like most Americans, I sat awestruck in front of the television on that summer night in 1969 when Neil Armstrong and Buzz Aldrin took humanity's first steps onto another celestial body. But perhaps the biggest key to my excitement was the potential this event embodied: Armstrong and Aldrin were both in their late 30s when they made their visit to the moon, and I had every reason to believe that, by the time I reached that age, I'd not only be following in their footsteps, but I'd be on my way to even bigger and better things.

I'm in my late 30s now, and we all know, of course, that none of this has happened. The reasons for this are numerous, and involve various political, financial, and technological factors; these have been discussed in numerous other forums, and I see no need to repeat those discussions here. But let me take a minute to point out the human side of this: I am one of numerous individuals in my generation who, after receiving the strong post- *Sputnik* math/science education and being stimulated by the *Apollo* project's missions, wanted nothing more than to grow up and make our own contribution to what we thought of as humanity's greatest endeavor. Now that we are ready to "step up to the plate," as it were, we find that the game has been called, and there is no place for us.

My experience is probably pretty typical, although I should point out that I took a couple of detours along the way to my Ph.D., and consequently I was somewhat older than most Ph.D. recipients. Like almost everybody else, I applied for numerous jobs after earning my degree, but apart from a temporary position I held for awhile at a nearby museum, my job search was entirely unsuccessful, to the point where I was never even able to obtain an interview. After a few dozen rejections like this, it was tempting to start believing that there was something wrong with me, but it only took a few discussions with other new Ph.D. recipients to show that this situation is the norm for young scientists in our society today. Almost every "decent" position – i.e., a permanent position at a university or a research lab – typically has at least 100 applicants, with incidents of 200, 400, even 800 or more applicants not being especially uncommon. True, there are the temporary positions – the "postdocs," as they're called, which usually evaporate after two or three years – and while a tour as a postdoc (which can be considered equivalent to a "residency" in the medical profession) is usually considered normal for a scientist's career, these days it is not at all unusual to encounter young scientists on their second, third, or higher-numbered postdoc, wondering if they'll ever be able to land a permanent position of employment "someday."

THE FUTURE THAT WASN'T. Buzz Aldrin standing on the Sea of Tranquility. NASA photograph.

To see why this is happening, consider that we are of the generation that was especially encouraged to go into science, and thus the proportion of that generation who chose to pursue a scientific career is perhaps somewhat higher than normal. At the same time, most of us were led to believe that the scientists and engineers hired in the wake of *Sputnik* would be nearing retirement at the time we came of age, and that we could then step in and continue on for them. As it turned out, this was half true; indeed, the preceding generation of scientists is retiring, but as they leave, their positions are leaving with them. Thus, we have a significantly greater number of individuals – all highly intelligent, highly educated, and highly motivated – competing for pieces of a significantly smaller pie.

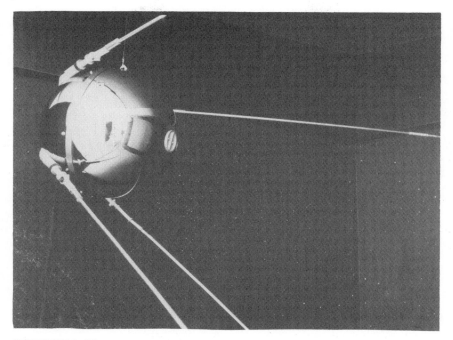

SPUTNIK 1. *The event which launched the Space Age, and led to a strong push on science education in America. Photograph provided courtesy of The Space Center, Alamogordo, New Mexico.*

When I earned my Ph.D. I had just turned 34, and had a wife and two small children. After about a year and a half of unsuccessfully hunting for a decent job, I could clearly see "the writing on the wall"; the best I could probably hope for was a succession of postdocs, until some undefinable point in the distant future when I could either land a permanent position, or eventually get disgusted and leave the field. If I were to choose the latter, of course, I could save myself a lot of trouble and leave the field before getting started. Neither of these alternatives appealed to me; I had invested too much of my efforts, my hopes, my dreams – my entire life – in this enterprise to simply just give it all up. At the same time, I was extremely reluctant to subject myself and my family to this prospect of forced uprootings and relocations every two years for the indefinite future. Thus, almost as an act of desperation, I decided to form my own organization for scientific research and education. Succeeding in this endeavor would be a longshot, and I knew that, but I decided that making this attempt,

131

even if it ended in failure, was infinitely preferable to sitting around and doing nothing except feeling sorry for myself.

MAD *MAGAZINE TELLS IT LIKE IT IS. Dave Berg's "The Lighter Side of".* Copyright 1993, E.C. Publications, Inc. All rights reserved. Used with permission.

When I formed the Southwest Institute for Space Research in 1993 I deliberately included a strong educational component within the Institute's mission. I did this for several reasons, the most important of which is that, by doing my part to improve scientific literacy in our society, I hoped to help reverse the societal trend which produced the employment situation that scientists in my generation are finding themselves in. At the same time, I did it for a more personal reason: I discovered during my graduate school days that I thoroughly enjoyed working in education, and sharing my knowledge and enthusiasm of space to all age groups, from elementary school children to adults. And I did it for a very practical reason – it is easier to solicit grants and donations for educational purposes than it is for pure research.

I started off with high hopes and aspirations, and scored a few successes during the next couple of years, i.e.; I managed to publish a few research papers, and I was able to get my name known a little in the local area by a few of the educational activities I was able to conduct. But these were more or less token successes contained within a larger sea of disappointments, and I quickly learned that getting money to operate wasn't as easy as I had hoped it might be. By mid-1995 it was beginning to look like my gamble wasn't going to pay off after all. This was a hard prospect for me to face, for it left me back where I started two years earlier: what do I do now? I hope it's not hard to see why I was beginning to suffer some bouts of depression as the summer of 1995 rolled on.

And then, of course, came that night in July... Already, the name recognition brought to me by the discovery of Hale-Bopp has helped me win grants and donations that I wasn't able to get earlier. I can't quite claim that the Institute is financially stable at this point, but with the money that is starting to come in I am beginning to be able to do a few things with the Institute that I couldn't do before. Some of the research and educational projects are starting to get underway, and I am hopeful that the day is not too far in the future when I can declare that the Institute can hold its own, and is operating in earnest.

It is Comet Hale-Bopp – a pure stroke of random luck – that is making all this possible for me, and hopefully my readers are able to understand just how important this object is to me. For almost my entire life, through the inevitable peaks and valleys that have occurred in my life just like in everyone else's, my love of the night sky has been a constant. My good friend David Levy, co-discoverer of Comet Shoemaker-Levy 9, once pointed out that "if you approach the sky on her terms, she will show you many great and wondrous things." And he was right: at that pivotal moment in time, I wasn't doing any important scientific observation, I wasn't hurrying to discover anything or analyze any data; all I was doing was quietly enjoying what the sky had to offer. And look what happened...

I'm obviously enjoying the notoriety that this discovery has brought to me, and I'm thoroughly grateful for the opportunities that are opening up both for me personally and for my Institute. So where do I go from here? My answer to this harkens back to my discussion at the end of Chapter 3; i.e., I believe Comet Hale-Bopp presents a wonderful opportunity – perhaps the best opportunity that's come along in many, many years – to introduce people to the wonders of the heavens, and to science in general. In so doing, I would like to think we can see a substantial improvement in the scientific literacy in our society.

I've been throwing around that phrase, "scientific literacy," quite a bit, and I'd like to take a little closer look at it. Essentially, a person who is scientifically literate has a sufficient background in basic science *and how science is done* so that he/she is equipped to be able to make decisions about how science should proceed. I'm not saying I need the public to be experts in all forms of science – that would be impossible – but informed enough about the current state of science so as to be able to make intelligent decisions about it. It is, after all, the public which supplies much of the funding that allows scientific research to proceed, and it is only right that the public has a major say in how that funding is used. But – and I'm speaking as a scientist here – if the public is going to decide which science should be pursued, I want to make sure that that public knows something about science.

For example, should our society engage in an active, vigorous, space program with the eventual goal of colonizing other planets? To me, the answer is "yes;" there are others, I'm sure, to whom the answer is "no." Since it is the public's money that would be supporting such an endeavor to a large degree, the public is entirely justified in asking me to defend my "yes" answer and in asking those who say "no" to defend their answer. I just want to make sure, though, that the public is sufficiently knowledgeable, in both the key specifics of the question and in the proper scientific principles, to understand the defense I would give.

Every American should learn enough... **ASTRONOMY**

...to ask themselves, "How can stars that are zillions of miles away know that I should 'avoid emotional entanglements tomorrow'?!"

As our society moves into the 21st Century the public is increasingly going to be asked to render its decisions on numerous scientific topics – about space development, about AIDS, about environmental issues, about biomedical research, and so on. How the public decides on these issues will drastically affect how our society develops over the next several decades, and will have a tremendous impact on future generations. It is thus imperative that the public understand these issues as fully as possible so that the most intelligent and informed – the "best" – decisions can be made.

Let me illustrate what I mean by "scientific literacy" with one somewhat tongue-in-cheek example. If the public as a whole were as scientifically literate about astronomy as I think it should be, I would not have had to write the "And the Misinformation Flies ... " section in Chapter 2. The world always has had and always will have those with "strange" ideas – for example, those of "The World Will End Tuesday" variety – and for the most part the public rightfully dismisses them. But let those individuals start dressing up their chatter with selectively chosen bits of science, and they begin leading members of the public astray. All the ideas about Hale-Bopp I discussed in that section are sheer nonsense, and would have been recognized by a scientifically literate person as such, but from my readings of the printed media and from the discussions on the Internet, and from some of the telephone calls and letters I have received, it is clear that at least some segments of the public have given these ideas considerably more credibility than they deserve. Following from this, what I'd like to see then is a scientifically literate public that would be able to recognize this kind of mininformation as the nonsense it is, and would be able to dispense with them right away and go on about the business of making the important decisions it has to make.

OK, I've said my piece about scientific literacy and I thank my readers for hearing me out. I think it's clear to everybody that I'm as excited as can be about the potential display of Comet Hale-Bopp. As someone who's been following comets for many years, I had always wanted to discover one and have my name on it someday, and now it's happened, and not just with any comet, but clearly a comet that's destined to go down in the history books. (What do I do for an encore? Well,...) People ask me how I feel about it, and I'm not sure I can give a decent answer to that. I'm tempted to answer with one of the old platitudes about "this can't be real, this must be a dream," but I'm not even sure that would be a true statement. As hard as it might seem to believe, this *is* real, and I do know it. One thing, at least, seems clear: this is almost certainly a once in a lifetime event, and I can be pretty sure that nothing like this will ever happen to me again. Thus, I intend to fully enjoy it...

To all my readers, thank you for buying this book; I hope this has been as enjoyable for you as it has been for me. Now, what I want you to do is go out to a dark site on an occasion or two during those first few months of 1997, look up, and enjoy my comet. Take a moment, and just appreciate the beauty that the heavens have to offer us. And then – and here is my parting shot – I want you to try to imagine the world that Hale-Bopp will see when it makes its next return in about 3400 years. Of course, there's no way we can realistically do this with any confidence. But keep this in mind: the world that we leave our descendants of that far-future era begins with us, here, today.

Cloudcroft, New Mexico
April 21, 1996

APPENDIX A: SOURCES OF INFORMATION

BOOKS

The following list of books currently available about comets can in no way be considered complete. These are some of the better books about comets that I am aware of, but there are certainly others.

GENERAL INFORMATION

I start here with books that are most accessible to the general reader, and work my way up to some of the more advanced works.

Comets: An Illustrated Introduction by Patrick Moore, Charles Scribner's Sons (New York), 1976. An engaging, highly readable description of comets from a well-known British popularizer of astronomy. Slightly dated at this point.

Comets: Vagabonds of Space by David A. Seargent, Doubleday & Company, Inc. (Garden City, New York), 1982. Another very readable account, slightly more in-depth than Moore's book, although also slightly dated. Seargent is an active comet observer who has discovered one comet.

Comets by Don Yeomans, John Wiley & Sons (New York), 1991. Highly engaging, detailed up-to-date treatment of comets from one of the world's top experts on the subject.

The Quest for Comets by David H. Levy, Avon Books (New York), 1995. Very readable and highly personal discussion of comets from a well-known amateur astronomer and discoverer or co-discoverer of numerous comets, including Comet Shoemaker-Levy 9.

The Mystery of Comets by Fred Whipple, Smithsonian Institution Press (Washington, D.C.), 1985. One of the earliest installments of the Smithsonian Library of the Solar System. A somewhat technical account of comets, from one of the top cometary astronomers of the 20th Century, and originator of the "dirty snowball" model of the cometary nucleus.

Rendezvous in Space: The Science of Comets by John C. Brandt and Robert D. Chapman, W.H. Freeman and Company (New York), 1992. Another somewhat technical treatment of comets, from two prominent cometary scientists. (These same two authors earlier had written *Introduction to Comets*, Cambridge University Press, 1981, which can perhaps be considered an earlier edition of *Rendezvous in Space.)*

Comets edited by Laurel L. Wilkening, University of Arizona Press (Tucson, AZ), 1982. A series of review articles about the current state of scientific cometary knowledge. Informative, but geared primarily toward professional astronomers and other advanced readers.

OBSERVING COMETS

International Halley Watch Amateur Observers' Manual for Scientific Comet Studies. Part I. Methods by Stephen J. Edberg, NASA–Jet Propulsion Laboratory (Pasadena, CA), 1983 (JPL Publication 83-16). Written for the participants of the International Halley Watch's Amateur Observation Network (of which Edberg was coordinator), this is a guide on how to obtain scientifically useful observations of comets. Although written primarily to guide in observations of Halley's Comet, the methods discussed are applicable to all comets. (Part II is an ephemeris for Halley's Comet during the period of its visibility.)

Observing Comets, Asteroids, Meteors, and the Zodiacal Light by Stephen J. Edberg and David H. Levy, Cambridge University Press, 1994. An observing manual for comets and other minor objects of the solar system, from two well-known and experienced amateur astronomers.

SPECIFIC INFORMATION

Comets: A Descriptive Catalog by Gary W. Kronk, Enslow Publishers (Hillside, N.J.), 1984. A description of each of the comets that had appeared up through mid-1982. (Much of the material on the earlier comets is taken from the more technical *Physical Characteristics of Comets* by Russian cometary astronomer S.K. Vsekhsvyatskii, published in Moscow in 1958, with English translation by the Israel Program for Scientific Translations, Ltd., 1964.) Fairly accessible to the layperson, although because of the publication date information on the most recent comets is necessarily missing. Kronk is presently working on a more in-depth work, *Cometography*, which should be in press soon.

Catalogue of Cometary Orbits by Brian G. Marsden and Gareth V. Williams, Minor Planet Center, Smithsonian Astrophysical Observatory (Cambridge, MA). A catalogue of orbital elements for all comets for which these have been determined. New editions are published on approximately an annual basis; the most recent edition, the 11th, was issued in January 1996.

FICTION

Lucifer's Hammer by Larry Niven and Jerry Pournelle, Ballantine Books (New York), 1977. A detailed and highly engaging account of a comet's collision with the earth, and the subsequent effects on society, from two highly acclaimed science fiction authors. Several individuals have pointed out that the particular comet in this story, Hamner-Brown, has the same initials as Hale-Bopp!

The Hammer of God by Arthur C. Clarke, paperback version Bantam Books (New York), 1994. A crisp and lucid account of the discovery of an asteroid on a direct collision course with the earth, and the attempts to divert its path, by one of the acclaimed "Grand Masters" of science fiction (author of, among many other works, *2001: A Space Odyssey*.) A motion picture of this novel, to be directed by Steven Speilberg, is currently in development.

Heart of the Comet by Gregory Benford and David Brin, Bantam Books (New York), 1986. About a colonizing expedition to Halley's Comet during its next return in 2061, from two well-known science fiction authors.

Isaac Asimov's Wonderful Worlds of Science Fiction #4: Comets edited by Isaac Asimov, Martin H. Greenberg and Charles G. Waugh, paperback version Signet Books (New York), 1986. A collection of 20 science fiction short stories, dating from the 17th Century up through 1985, illustrating some of the effects comets have had on the human psyche during the past several centuries.

In the Days of the Comet by H.G. Wells (originally published in 1906). An "early" story of a comet's impacting the earth, from the classic British writer of science fiction.

PERIODICALS

The following two monthly magazines are excellent guides to the universe around us, and can be guaranteed to provide up-to-date information on Comet Hale-Bopp. Both are usually available in popular newsstands and bookstores.

Astronomy, published by Kalmbach Publishing Co., 21027 Crossroads Circle, P.O. Box 1612, Waukesha, WI 53187.

Sky & Telescope, published by Sky Publishing Co., 49 Bay State Road, Cambridge, MA, 02138.

More technically-oriented readers may be interested in reading the *International Comet Quarterly*, available by subscription from the Smithsonian Astrophysical Observatory, 60 Garden Street, Cambridge, MA, 02138. This journal contains papers on comets and comet observing from amateur and professional astronomers, information on currently visible comets, and an archive of cometary brightness measurements.

Announcements of new discoveries of comets and other astronomical objects, as well as a lot of other information of transient astronomical phenomena, are given in the *IAU Circulars*, issued irregularly (i.e., when there's a need to), and available via subscription from the Smithsonian Astrophysical Observatory. Subscribers can receive the *Circulars* on printed postcards, and/or electronically via modem and the Internet, and via email.

ON-LINE INFORMATION

As with the books, I make no pretense that the following list of on-line information sites is complete with regard to what's available for Comet Hale-Bopp. These are the information sites that I am aware of, and readers who examine these will probably come across others.

USENET

Specific information on Comet Hale-Bopp results is usually posted to the USENET newsgroup **sci.space.news** as developments occur. I've seen discussions about the comet in the groups **sci.astro, sci.astro.amateur** and **sci.space.science**. It's fair to warn the reader that many of these discussions center around the misinformation I discussed in Chapter 2.

TELNET

The University of Maryland's Hale-Bopp bulletin board can be accessed via TELNETat **pdssbn. astro.umd.edu**; the username is **halebopp** and no password is required. This account was set up primarily for use by comet scientists, so much of the discussion is rather technical. Be aware that this is a UNIX-based account.

FTP

Images of Comet Hale-Bopp (and many other objects) taken with the *Hubble Space Telescope* can be acquired via anonymous FTP from the FTP site **ftp.stsci.edu**. Images in GIF format are in the directory **/pubinfo/gif** and in JPG format in the directory **/pubinfo/jpeg**.

WORLD WIDE WEB

Again, please note that at best this is only a partial list. The following URL sites are more or less devoted exclusively to Comet Hale-Bopp:

http://www.halebopp.com – "The Hale-Bopp homepage," a web site geared toward being accessible to the layperson. This site includes contributions from both myself and Tom Bopp.

THE FRONT COVER OF www.halebopp.com. Courtesy Russell Sipe and Sipe Electronic Publishing.

http://www.comet.arc.nasa.gov/comet/ – This site hosted the "Night of the Comet" "virtual star party" that was conducted during the passage of Comet Hyakutake in March 1996. The "Hale-Bopp: Live in the American Classroom" project (see Appendix D) will be operated in conjunction with this homepage during the period of the comet's naked-eye visibility.

http://NewProducts.jpl.nasa.gov/comet/ – Maintained at JPL by Ron Baalke, moderator of the USENET newsgroup **sci.space.news.**

http://medicine.wustl.edu/˜kronkg/1995_O1.html – Maintained by Gary Kronk (author of *Comets: A Descriptive Catalog* and *Cometography* – see above).

http://www.eso.org/comet-hale-bopp/comet-hale-bopp.html – Maintained by the European Southern Observatory, located in Chile and headquartered in Switzerland. Includes many of the latest images and scientific results.

http://www.indirect.com/www/polakis/halebopp.html – A planetarium-generated "movie" of Hale-Bopp's motion through the inner solar system.

The following URL sites are more general, containing information about many other comets and objects in addition to Hale-Bopp:

http://encke.jpl.nasa.gov – This is the "comets" homepage, maintained at JPL by Charles Morris, one of the world's most active comet observers. In addition to images and information about many comets, this site includes the excellent summary "Information on Comet Hale-Bopp for the Non-Astronomer."

http://cfa-www.harvard.edu/cfa/ps/icq.html – The homepage for the *International Comet Quarterly* (see above).

http://pdc.jpl.nasa.gov/stardust/ - The homepage for the *Stardust* mission (see pp. 123-124_

http://seds.lpl.arizona.edu/billa/tnp/comets.html – General information about comets, and links to several other sites for more specific information.

http://fly.hiwaay.net/˜cwbol/astron/comet.html – A directory of available cometary images and information.

http://pdssbn.astro.umd.edu – The URL for the University of Maryland's various transient phenomena bulletin boards (of which Hale-Bopp is one among several).

http://www.stsci.edu/pubinfo/Pictures.html – The URL for images obtained with the *Hubble Space Telescope*.

http://www.skypub.com – Homepage for *Sky & Telescope* magazine.

http://www.kalmbach.com/astro/astronomy.html – Homepage for *Astronomy* magazine.

MISCELLANEOUS INFORMATION

PHOTOGRAPHY OF COMETS

Chapter 6 of Steve *Edberg's International Halley Watch Amateur Observers' Manual* . . . (see above) discusses some of the generalities involved in comet photography, and contains an extensive bibliography for investigating some of the many specifics. Many practical suggestions are given in the articles "Catch a Comet on Film" by Rick Dilsizian in the January 1996 issue of *Astronomy* (p. 78) and "Comet Photography for Everyone" by Dennis di Cicco and Leif J. Robinson in the May 1996 issue of *Sky & Telescope* (p. 28).

CCD IMAGING

The quarterly magazine *CCD Astronomy*, published by Sky Publishing Co., 49 Bay State Road, Cambridge, MA, 02138 gives regular updates and ideas on taking astronomical photographs with CCDs.

COLLISION THREATS

An excellent summary of this subject is given in the article "Target: Earth" by David Morrison in the October 1995 issue of *Astronomy* (p. 34). *Rogue Asteroids and Doomsday Comets* by Duncan Steel (Wiley, New York, 1995) is somewhat more mathematical but still is geared toward a general audience. More technical discussions are given in *Hazards Due to Comets and Asteroids*, edited by Tom Gehrels, University of Arizona Press (Tucson), 1995 and in the NASA Solar System Exploration Division's *Report of the Near-Earth Objects Survey Working Group* by Eugene Shoemaker *et al.,* June 1995.

Two of the more worthwhile books on the Shoemaker-Levy impacts with Jupiter are *The Great Comet Crash: Comet Shoemaker-Levy 9 and its Impact on Jupiter,* edited by John R. Spencer and Jacqueline Mitton (Cambridge University Press, 1995) and *Impact Jupiter: The Crash of Comet Shoemaker-Levy 9* by David H. Levy (Plenum Press, 1995). The latter book, written by one of the comet's co-discoverers, offers an insider's view of the events surrounding its discovery and the eventual realization of its catastrophic fate.

THE FRONT COVER OF ASTRONOMY, *OCTOBER 1995 ISSUE. Copyright 1995 Kalmbach Publishing Co., and* ASTRONOMY *magazine. Reproduced with permission.*

ECLIPSES

Detailed guides for both the March 9, 1997 and February 26, 1998 total solar eclipses, prepared by Fred Espenak (NASA Goddard Space Flight Center, Greenbelt, MD) and Jay Anderson (Environment Canada, Winnipeg, Manitoba) are available from the NASA Center for AeroSpace Information, 800 Elkridge Landing Rd., Linthicum Heights, MD 21090-2934, phone (301) 621-0390. The appropriate publication numbers are NASA Reference Publication 1369 for the 1997 eclipse and 1383 for the 1998 eclipse.

The only tours to the 1997 eclipse that I'm aware of are those offered by Boojum Expeditions, 14543 Kelly Canyon Rd., Bozeman MT 59715, phone (406) 587-0125; by Explorers Tours, 223 Coppermill Rd., Wraysbury TW19 5NW, England; and by the Astronomical League, Rt. 2, Box 940, Bartlesville, OK 74006, phone (918) 333-1966 (contact person Ken Willcox).

For the 1998 eclipse Tom Bopp and I will be traveling with Tropical Adventures, 2516 Sea Palm Dr., El Paso, TX 79936, phone (800) 595-1003. Other tours that I'm aware of include those being offered by Scientific Expeditions, Inc., 227 West Miami Ave., Suite #3, Venice, FL 34285, phone (800) 344-6867; by Twilight Tours, Inc., 3316 W. Chandler Blvd., Burbank, CA 91501-2511, phone (818) 841-8245; and by the Astronomical League (address above).

These lists are almost certainly not complete, and various other tours will probably become available as the respective dates approach. Many of these tours will probably advertise in *Sky & Telescope* and *Astronomy,* and interested readers may want to scan those magazines.

COMET-WATCHING TOURS

During the early spring of 1997 Comet Hale-Bopp will remain above the horizon all night from latitudes far enough north. The Planetary Society (65 North Carolina Ave., Pasadena, CA 91106-2301, phone (818) 793-5100, is putting together a land tour package, probably to Alaska, during this time. (I will probably be going with The Planetary Society on this tour.)

Scientific Expeditions (see address above) is planning a comet watching cruise to the Caribbean at the end of March. The Planetary Society is also tentatively planning a tour to Hawaii at about this same time. Note that, in a sense, these are going in the wrong direction; the further *north* one is, the higher the comet will be located in the sky. There are, of course, other considerations...

Tropical Adventures (see address above) is tentatively offering a cruise to the Caribbean islands in early June 1997 (at which time the comet will be too far south for viewing from the continental U.S.)

As with the eclipse tours above, other similar tours may be offered during the coming months, and interested readers should be on the lookout for the appropriate ads in *Sky & Telescope* and *Astronomy*.

LIGHT POLLUTION

The International Dark-Sky Association (IDA) is an organization devoted to controlling unnecessary artificial lighting and in restoring the natural darkness and beauty to the night sky. IDA can be contacted at 3545 N. Stewart, Tucson, AZ 85716, and via its World Wide Web homepage at **http://www. darksky.org/˜ida**. The New England Light Pollution Advisory Group (NELPAG) has recently been formed to work towards appropriate lighting control in the northeastern U.S.; interested readers should contact Dan Green at the Smithsonian Astrophysical Observatory, 60 Garden Street, Cambridge, MA 02138, or visit NELPAG's homepage at **http://cfa-www.harvard. edu/cfa/ps/nelpag.html**.

EQUIPMENT

The following manufacturers are among those I consider "reliable," although this is not necessarily a complete list. Many of these operate through equipment dealers scattered around the country (and the world); for the dealer nearest you, contact the manufacturer in question. Also, check out the various advertisements in *Sky & Telescope* and *Astronomy*, both for appropriate dealers and for other equipment that might be desirable.

TELESCOPES

Meade Instruments Corporation – 16542 Millikan Ave., Irvine, CA 92714, phone (714) 756-2291. (The telescope with which I discovered Comet Hale-Bopp was a Meade Instruments DS-16.)

Celestron International – 2835 Columbia St., Torrance, CA 90503, phone (310) 328-9560.

Orion Telescope Center – P.O. Box 1815, Santa Cruz, CA 95061-1815, phone (408) 464-5710. (Orion manufactures and sells a lot of accessories and equipment in addition to telescopes.)

BINOCULARS

Many of the dealers associated with the above manufacturers also sell good binoculars. Two of the more reliable brand names for binoculars are:

Bushnell Sports Optics Worldwide – 9200 Cody, Overland Park, KS, phone (913) 752-3400.

Fujinon – 10 High Point Dr., Wayne, NJ, phone (201) 633-5600.

CCDs

Santa Barbara Instrument Group – 1492 East Valley Road, #33, Santa Barbara, CA 93150, phone (805) 969-1851.

Meade Instruments Corporation (see address above) has also recently introduced a line of CCDs.

APPENDIX B: A HALE-BOPP TIMELINE

1993

April 27 – Earliest pre-discovery photograph (Rob McNaught, Siding Spring Observatory, New South Wales), 13.1 AU from sun

1995

May 29 – Earliest 1995 pre-discovery photograph (Terrence Dickinson, Chiricahua Mountains, Arizona), 7.6 AU from sun

July 5 – opposition

July 23 – discovery by Alan Hale and Thomas Bopp; 7.15 AU from sun

July 26 – announcement of preliminary orbit; official naming as Comet Hale-Bopp

August 1 – announcement of "correct" orbit

August 10 – 7.0 AU from sun

September 26 – first Hale-Bopp photographs taken with *Hubble Space Telescope*

October 23 – second set of Hale-Bopp photographs taken with *Hubble Space Telescope*

November 23 – last visual sighting before conjunction with sun (Alan Hale, Cloudcroft, New Mexico)

December 2 – 6.0 AU from sun

December 8 – last photographs taken before conjunction with sun (Siding Spring Observatory, New South Wales)

December 10 – last radio observations before conjunction with sun (Kitt Peak National Observatory, Arizona)

1996

January 3 – conjunction with sun

February 1 – first photograph taken after conjunction with sun (Gordon Garradd, Loomberah, New South Wales)

February 2 – first visual sightings after conjunction with sun (Arturo Gomez, Cerro Tololo Interamerican Observatory, Chile, and Terry Lovejoy, Jimboomba, Queensland)

February 28 – Hale-Bopp crosses plane of Earth's orbit from south to north, 5.17 AU from sun

March 17 – 5.0 AU from sun

April 5 – closest approach to Jupiter (0.77 AU)

April 7 – most recent *Hubble Space Telescope* observations
May 6 – 4.5 AU from sun
May 8 – occultation of Hale-Bopp by the moon
May 10 – beginning of International Comet Hale-Bopp Days, period 1
May 18 – first naked-eye sighting (Steve O'Meara, Volcano, Hawaii)
May 31 – end of International Comet Hale-Bopp Days, period 1
June 23 – 4.0 AU from sun
July 1 – full moon
July 3 – opposition
July 7-23 – best times for viewing
July 15 – new moon
July 30 – full moon
August 5-20 – best times for viewing
August 8 – 3.5 AU from sun; water sublimation and increase in brightness?
August 14 – new moon
August 28 – full moon
September 1-15 – best times for viewing
September 12 – new moon
September 21 – 3.0 AU from sun
September 26 – full moon; total eclipse of moon (totality 8:18 PM to 9:29 PM
     MDT)
September 29 – beginning of International Comet Hale-Bopp Days, period 2
October – easily visible to naked eye from dark site?
October 1-15 – best times for viewing
October 12 – new moon
October 19 – end of International Comet Hale-Bopp Days, period 2
October 26 – full moon
October 30–November 13 – best times for viewing
November 2 – 2.5 AU from sun
November 10 – new moon
November 24 – full moon
November 27–December 12 – best times for viewing; comet low in west during
     and after dusk
December 10 – new moon
December 12 – 2.0 AU from sun
December 15 – beyond this point, comet probably too close to sun to be seen
     easily, except in far northern latitudes
December 24 – full moon
December 31 – conjunction with sun

1997

January 3 – possible meteor shower associated with Hale-Bopp

January 7-22 – best times for viewing; comet low in northeast morning sky, but probably fairly bright

January 9 – new moon

January 14 – Hale-Bopp at highest point north of Earth's orbital plane (1.20 AU)

January 21 – 1.5 AU from sun

January 23 – full moon

January 27 – Hale-Bopp located 7° northwest of bright star Altair

February – Hale-Bopp easily visible to naked eye (hopefully!)

February 4-20 – best times for viewing

February 7 – new moon

February 15–March 1 – highest above northeastern horizon before dawn (moonlight will interfere during latter portion of this)

February 22 – full moon

March – Hale-Bopp spectacular to naked eye in pre-dawn sky?

March 3 – beginning of International Comet Hale-Bopp Days, period 3

March 4 – Hale-Bopp directly "above" sun, 1.04 AU from it

March 6-21 – best times for viewing in morning sky

March 9 – new moon; total solar eclipse visible from Mongolia and Siberia (comet should be easily visible during totality)

March 9 – 1.0 AU from sun

March 20-30 – comet visible all night from north of latitude 45°; for rest of northern hemisphere, comet visible both in evening and morning skies

March 22 – closest approach to Earth (1.32 AU)

March 22 – conjunction with sun

March 23 – full moon; partial lunar eclipse (midpoint 9:40 PM MST)

March 25 – comet at most northern point in its path across the sky

March 25 – Hale-Bopp located 5° north of Andromeda Galaxy

March 26–April 10 – best times for viewing in evening sky; spectacular to naked eye (hopefully!)

April 1 – perihelion, 0.91 AU from sun

April 7 – new moon

April 10 (approximately) – near highest point above horizon in evening sky

April 12 – end of International Comet Hale-Bopp Days, period 3

April 22 – full moon

April 24 – 1.0 AU from sun

April 25–May 10 – best times for viewing (last good views from most of northern hemisphere)

May 6 – Hale-Bopp crosses plane of Earth's orbit from north to south, 1.11 AU from sun

May 6 – new moon

May 6 – Hale-Bopp located 10° northeast of Hyades star cluster in Taurus

May 20-31 – last sightings from mid-northern latitudes

May 21 – full moon

May 24–June 9 – best times for viewing; easily visible from southern hemisphere, and possibly still quite spectacular

June 5 – new moon

June 5 – Hale-Bopp located 2° northeast of bright star Betelgeuse in Orion

June 10 (approximately) – comet sets at sunset from mid-northern latitudes

June 10 – 1.5 AU from sun

June 15 – Hale-Bopp located 3½° west of Encke's Comet; the latter object should be visible in front of Hale-Bopp's tail around this date

June 20 – full moon

June 23–July 2 – best times for viewing (southern hemisphere only; comet visible low in northwest during and after dusk)

July 3 – conjunction with sun

July 4 – new moon

July 4-18 – best times for viewing (southern hemisphere only; comet visible in east during and before dawn); still bright to naked eye?

July 20 – 2.0 AU from sun

July 20 – full moon

July 24 – Hale-Bopp located 9° northeast of bright star Sirius

August 1-16 – best times for viewing (southern hemisphere only)

August 3 – new moon

August 18 – full moon

August 26 – beginning of International Comet Hale-Bopp Days, period 4

August 29 – 2.5 AU from sun

August 30–September 14 – best times for viewing (southern hemisphere only); faintly visible to naked eye?

September 1 – new moon

September 10 (approximately) – comet rises about beginning of dawn from mid-northern latitudes; visible low in southeast

September 16 – full moon; total lunar eclipse visible from eastern hemisphere

September 16 – end of International Comet Hale-Bopp Days, period 4

September 29–October 13 – best times for viewing; best opportunities for final observations from mid-northern latitudes

October 1 – new moon

October 10 – 3.0 AU from sun

October 15 – full moon

October 28–November 12 – best times for viewing (southern hemisphere); last sightings with naked eye? Last potential opportunity for observations from mid-northern latitudes

October 31 – new moon

November 12 (approximately) – last possible observations from mid-northern latitudes

November 14 – full moon

November 14 (approximately) – beyond this date, visible above horizon all night from mid-southern latitudes

November 23 – 3.5 AU from sun

November 30 – new moon

December 14 – full moon

December 27 – opposition

December 29 – new moon

1998

January 3 – possible meteor shower associated with Hale-Bopp

January 8 – 4.0 AU from sun

February 25 – 4.5 AU from sun

February 26 – total solar eclipse, visible from northern South America and Caribbean Sea

February 26 (approximately) – last probable sighting of Hale-Bopp by Hale and Bopp; from Caribbean islands comet visible low in the southwest during evening hours

April 17 – 5.0 AU from sun

June 25 – conjunction with sun (well south of sun, still visible all night from mid-southern latitudes)

July 10 – Hale-Bopp located 1° northeast of bright star Canopus

August 1 – 6.0 AU from sun

November 23 – 7.0 AU from sun

December 11 – 7.15 AU from sun; same distance as at discovery

December 24 – opposition

1999

January–June – last visual sightings by amateur astronomers?

January 3 – possible meteor shower associated with Hale-Bopp

March 24 – 8.0 AU from sun

June 22 – conjunction with sun

July 31 – 9.0 AU from sun

December 13 – 10.0 AU from sun

December 20 – opposition

2000

May–June – Hale-Bopp passes directly across central region of Large Magellanic Cloud
May 3 – 11.0 AU from sun
June 9 – Hale-Bopp almost directly over location of 1987 supernova in Large Magellanic Cloud
September 28 – 12.0 AU from sun
December 15 – opposition

2001

March 11 – 13.1 from sun; same distance as at April 1993 pre-discovery photograph

2003

December 1 – 18.8 AU from sun; same distance as Halley's Comet at time it was last photographed
December 4 – opposition

APPENDIX C: GLOSSARY OF TECHNICAL TERMS

Aphelion: that point in an object's orbit which is farthest from the sun.

Asteroid belt: the region of the solar system, between the orbits of Mars and Jupiter, wherein the majority of the asteroids orbit the sun.

Astronomical Unit (AU): a unit of distance equivalent to the average distance between the earth and the sun (93 million miles, or 149 million km). The distances of objects within the solar system are usually expressed in AU.

Central condensation: a dense cloud of material surrounding a comet's nucleus, and which usually appears as a bright spot within the coma.

Charge Coupled Devices (CCDs): electronic imaging systems which utilize a computer chip to convert received light into electrical charges, and then store this information as a computer-readable image. CCDs can record faint objects within a small fraction of the exposure time normally required for ordinary photographic film, and are rapidly becoming the "cameras of choice" for both professional and amateur astronomers.

Coma: the "fuzzy" head of a comet, composed of gas and dust which has been ejected off the nucleus as it approaches the sun.

Conjunction: literally, a meeting or close gathering of two or more objects. When applied to astronomical objects, we usually mean that the two objects are located along the same line of sight as seen from the earth. A comet in conjunction with the sun is located along the same line of sight with the sun and thus is usually invisible. However, if a comet's orbit is highly inclined with respect to the earth's orbit (see "inclination" below), the comet may be located well to the north or south of the sun at the time of their conjunction, and thus may still be visible from Earth (although usually with difficulty).

Corona: the faint, tenuous (but hot) outer atmosphere of the sun, usually visible only during a total solar eclipse.

Deep-sky object: distant objects, other than individual stars, located well beyond the solar system. The term usually encompasses star clusters, gas and dust clouds in space, and other galaxies outside of the Milky Way.

Degree (°): a unit of angular measure, equal to 1/360 of a full circle. Sizes of objects in the sky, and distances between them, are usually expressed in degrees. The full moon is approximately ½° in diameter; a fist at arm's length is approximately 10° tall; the distance between the horizon and the zenith is ¼ of a circle, or 90°, etc.

"Dirty Snowball": the theory originally put forth by Fred Whipple in 1950 concerning the structure of a comet's nucleus, which was verified when the spacecraft *Giotto* flew by the nucleus of Halley's Comet in 1986.

Ecliptic: the path across the constellations upon which the sun travels.

Elements: the mathematical quantities which define the parameters of an object's orbit around another object, for example, of a comet's orbit around the sun. These elements include the date and location of perihelion, the inclination of its orbit with respect to the earth's orbit, etc.

Ellipse: an oval-shaped closed curve. The orbits of all planets, asteroids, and (most) comets around the sun are ellipses. The ellipses within which the planets and most of the asteroids travel are nearly circular, whereas the comets travel in orbits which are far more "elliptical."

Ephemeris (plural *ephemerides*): a table listing an object's position in the sky on different dates. The ephemeris for an object like a comet is computed from its orbital elements.

Hyperbola: a large open-ended curve; an object traveling in a hyperbolic orbit will never return to the object it is orbiting.

Inclination: a measure of how much an orbit is tilted with respect to the earth's orbit, usually expressed in degrees. An object moving in the exact same plane as the earth's orbit has an inclination of 0°, and an object moving in an orbit perpendicular to the earth's orbit has an inclination of 90°. An orbit with an inclination greater than 90° is said to be "retrograde," i.e., moving in a direction opposite that of the earth.

Indeterminate: when applied to orbits, a term which means that no definitive orbit can be determined. Usually this means that a wide range of orbits can fit the available data, and it is not possible to distinguish which of these orbits is the "correct" one.

Intrinsic brightness: when referring to a comet, the brightness that that comet exhibits as a result of its own size and activity, without regard for any external factors, such as its distance from the sun or the earth.

Ionized: electrically charged. (More correctly, an atom which has been ionized either has had negatively-charged electrons stripped from it, resulting in a positive ion, or has extra electrons added to it, resulting in a negative ion.)

Jet: an emission of material off a comet's nucleus, similar to the eruption of a geyser.

Jupiter-family comet: a comet which has been "captured" by Jupiter's gravity into a very small short-period orbit. Most Jupiter-family comets have orbital period of 6 to 8 years.

Kuiper Belt: a broad band of comets circling the sun at distances of a few hundred AU.

Large Magellanic Cloud: one of two small satellite galaxies of the Milky Way galaxy (the other object being called the "Small Magellanic Cloud"). Both of the Magellanic Clouds appear in the sky as separate detached portions of the Milky Way; neither are visible from mid-northern latitudes.

Light pollution: the unnecessary brightening up of the night sky through uncontrolled usage of artificial light.

Meteor shower: those occasions when the earth intersects a stream of interplanetary dust particles, creating the effect that numerous meteors appear to emanate from the same location in the sky.

Non-gravitational forces: small changes in a comet's orbital motion, caused by the eruptions of material off the nucleus. These eruptions act as small rocket engines, and thus can cause the comet to deviate slightly from its predicted course. Once a comet has been observed long enough, this effect can be included into the calculations of its orbit.

Nucleus: the "solid" object in the center of a comet. A comet's nucleus is composed primarily of dust mixed in with various ices, and when it approaches the sun, the sun's heat causes the ices to sublimate, creating the coma and the tail.

Occultation: an event wherein one solar system object (e.g., the moon or a planet) passes in front of a more distant object, as seen from our vantage point on the earth.

Oort Cloud: a large, spherical cloud of comets believed to enshroud the solar system at distances of 1000 to 10,000 AU.

Outburst: a sudden, dramatic increase in a comet's brightness (sometimes called a "flare").

Opposition: located in a position opposite that of the sun. A solar system object at opposition will usually rise around sunset, be at its highest point above the horizon around midnight, and will set around sunrise.

Parabola: an open-ended curve; an object traveling in a parabolic orbit will never return to the object it is orbiting. Most longer-period comets have orbits which are close to being parabolas.

Parallax: the apparent change in an object's position, as seen with respect to more distant objects, when observed from slightly different vantage points. Since there is a strict mathematical relationship between the size of an object's parallax and its distance, parallax is often used to find distances to astronomical objects.

Perihelion: that point in an object's orbit which is closest to the sun.

Periodic comet: see "Short-period comet" below.

Planetesimal: one of the "building blocks" that helped form the planets during the early stages of the solar system. Leftover planetesimals became the objects we know today as comets and asteroids.

Pre-discovery (photograph): a photograph of a comet (or other object) taken before that object's discovery, but not noticed until after the discovery has been reported. Often these images are found as a result of a deliberate search after an object's orbit has been computed.

Refraction: the "bending" of light as it passes through different substances (e.g., water, or a glass lens) or from one substance to another (e.g., from air to a vacuum). Refracting telescopes utilize this phenomenon through a lens in order to operate (as opposed to a "reflecting" telescope, which uses a mirror).

Retrograde: moving in a direction opposite that of the earth (see "inclination" above).

Scientific literacy: a condition wherein a person has enough knowledge of basic scientific facts and of overall scientific processes that he/she can make competent judgements of newly reported discoveries and can make informed and intelligent recommendations and decisions as to the future directions of scientific research.

Solar wind: a stream of highly energetic charged particles (such as protons, electrons, and ions) constantly "blowing" off the sun's surface.

Short-period comet: by definition, a comet with an orbital period of 200 years or less.

Star cluster: a close grouping of stars. Usually these are true associations of stars, containing anywhere from a dozen up through as many as a million individual stars, all located fairly closely to each other and traveling through space together.

Sublimate: to change directly from a solid to a gas. Substances such as water ice and carbon monoxide ice do this in space when they are heated, since the vacuum around them does not "force" them into a liquid state.

Sungrazer: a comet with a perihelion distance extremely close to the sun, usually less than 0.01 AU. Most of the known sungrazers travel in orbits remarkably similar to each other, and are thus believed to be the pieces of one large comet which split up a few thousand years ago.

Supernova: an exploding star; one which literally "blows itself to pieces." The last recorded supernova in the Milky Way appeared in the early 17th Century, but several are observed every year in other galaxies. The brightest supernova seen in the past several centuries appeared in 1987 in the Large Magellanic Cloud.

Zenith: the point in the sky directly overhead.

Zodiac: the group of constellations within which the sun, moon and planets travel.

APPENDIX D: ABOUT THE SOUTHWEST INSTITUTE
FOR SPACE RESEARCH

I formed the Southwest Institute for Space Research in 1993 to accomplish two main goals: *1)* to provide an environment for the performance of research in astronomy and space science; and *2)* to promote the scientific literacy of the public. I strongly believe that this latter goal can be accomplished most strongly by directly involving the public in the research endeavors of the Institute; thus, any research projects upon which the Institute embarks are specifically designed to include direct participation by members of the public, including both school students and laypersons, and participation of this kind is sought out and encouraged. In addition to this work I, along with any other scientists who might be attached to the Institute in the future, participate in the science education of our school children through classroom presentations, nighttime observing sessions, teacher training workshops, and so on.

I am currently in the process of initiating research efforts in several areas which are consistently among those considered fascinating and important by scientist and layperson alike. As I discussed at some length in Chapter 1, it has only been within the past two decades or so that we've realized there are objects in our solar system which can threaten our Earth, with potentially catastrophic effects should one of the larger objects collide with us. Many Earth-approaching objects have been discovered in recent years as a result of a few dedicated searches, but it has been estimated that less than 10% of the so-called "near-Earth" asteroids have been discovered so far. Here at the Institute we are in the initial stages of acquiring a research-grade telescope to be utilized in searching for some of that other 90%, and to assist in obtaining follow-up observations of objects that might be discovered elsewhere. (It has happened on several occasions that an object is discovered at an observatory somewhere, but poor weather on subsequent nights has precluded the necessary follow-up observations which are needed for a valid orbit to be determined.) We hope to have this program operational by the end of the decade.

One area which has held particular fascination to me and many other individuals throughout history is the question of whether or not other planetary systems, like ours, exist in the universe. In particular, we want to know if there are other planets like Earth "out there" and, if so, whether they might harbor any forms of life. Until very recently, the answers to questions of this nature were of necessity little more than speculation, but recent discoveries of planetary-sized companions to some nearby stars are beginning to tell us that planetary systems of some sort might exist in profusion after all. Some of the objects that have been found so far are somewhat unlike the objects that might be expected, but this, after all, is part of what makes this pursuit so fascinating. Surely many other discoveries of this nature lie ahead.

THE SEARCH FOR NEAR-EARTH ASTEROIDS. On May 19, 1996, a newly-discovered asteroid, 1996 JA$_1$, whizzed by only 280,000 miles from the earth (just outside the orbit of the moon). With a diameter of approximately 1300 feet, 1996 JA$_1$ could have destroyed a city if it had struck Earth in the wrong place. On this photograph of 1996 JA$_1$ taken four hours before its closest approach, the asteroid's motion causes its image to form a trail against the background stars, despite an exposure time of only 30 seconds. Photograph taken by Takuo Kojima, used with permission.

Along these lines, we at the Institute are initiating a program to acquire and archive data for those stars nearby to our solar system which are most like our sun. These stars are perhaps the most logical places to begin searches for planets, since the one star we do know that is accompanied by an Earthlike planet is, of course, the sun. "Sunlike" stars are actually fairly common, constituting about 15% of all the stars in our galaxy; several hundred of them are relatively "nearby" to the solar system, and for many of these objects very little is known. The program we're initiating is meant to rectify this, so that the appropriate information might be available when the time comes that full-scale planetary searches can begin in earnest.

A GLIMPSE OF THE FUTURE? The McDonnell Douglas DC-X (Delta Clipper - Experimental), an experimental single-stage to orbit (SSTO) rocket that has undergone several successful test flights at White Sands Missile Range in New Mexico. The DC-X and its successors may presage a future wherein access to space is routine and inexpensive. Courtesy McDonnell Douglas Aerospace and William Gaubatz.

Being a somewhat frustrated child of the *Apollo* area, I believe it is important to work on bringing our society to the point where we can begin to think seriously about becoming a true space-faring civilization. Accordingly, the Institute and I have been working with various representatives of industry and the education community to help bring about the development of an active commercial space program, and to assist in educating the public to the benefits to be derived from this. The bottom line is the oft-repeated phrase of "cheap access to space," and certainly reducing the costs of getting equipment and people to space is a necessity if we are to truly advance our knowledge of our space environment, and someday be able to leave this planet if we choose to. To cite one example, many of the planetary searches programs I mentioned above will have to be performed from above our atmosphere – either from low Earth orbit or, better yet, the moon – in order to produce significant results, and this is unlikely to occur until we can drastically reduce the costs of getting the equipment up there in the first place.

For the immediate future, though, we at the Institute are concentrating on projects to share the excitement of Comet Hale-Bopp with as broad an audience as possible. In particular, our project entitled "Hale-Bopp: Live in the American Classroom" has the ambitious goal of providing images of and information about the comet to every school, science museum, and planetarium in the country via the Internet and the World Wide Web. This project will be conducted in collaboration with the NASA Ames Research Center, and interested readers should be able to access the information through the "Night of the Comet" Web site given in Appendix A. Through this project, we can use one of the wonders of the nighttime sky to help bring the excitement of space to the current and next generations.

The Institute was incorporated within the state of New Mexico in December 1993, and received tax-exempt status from the Internal Revenue Service as a 501(c)(3) organization in August 1994. All contributions to the Institute are thus fully tax-deductible and are, of course, gratefully appreciated. By special arrangement with the publisher of this book, a percentage of the proceeds of every copy that is sold will go to the Institute so that it can carry out some of the projects I've discussed here.

ABOUT THE AUTHOR

Alan Hale was born in 1958 in Tachikawa, Japan, the son of a U.S. Air Force officer. His family moved to Alamogordo, New Mexico when he was a few months old, and Alan lived there until his graduation from high school in 1976. Afterward he attended the U.S. Naval Academy in Annapolis, Maryland, graduating with a Bachelor's Degree in Physics and a Naval officer's commission in 1980. After being stationed at assignments in San Diego and Long Beach, California, Alan left the Navy and worked for 2½ years as a Deep Space Network engineer at the Jet Propulsion Laboratory in Pasadena. He returned to New Mexico in 1986, entering graduate school at New Mexico State University in Las Cruces, and earned his Ph.D. in Astronomy in 1992. He founded the Southwest Institute for Space Research in southern New Mexico in 1993, and presently serves as its Director.

Alan has had a lifelong interest in astronomy and space, and acquired his first telescope at the age of 11. He has been an active visual observer of comets ever since, and has now observed over 200 of these objects up to the present time; this interest reached its culmination with his independent discovery of Comet Hale-Bopp in July 1995. In his professional life Alan primarily studies stars like the sun, and the potential they might have for planets and life, and has published several research papers in this field. He also writes a weekly newspaper column on space as well as occasional articles for science magazines and other periodicals.

Alan resides in Cloudcroft, New Mexico with his wife, Eva, and sons Zachary and Tyler. When not engaged in his astronomical pursuits he enjoys reading, hiking in the mountains, and watching football.

Photo of the author courtesy Marilyn Haddrill